LED Lighting

Sal Cangeloso

O'REILLY®
Beijing • Cambridge • Farnham • Köln • Sebastopol • Tokyo

LED Lighting

by Sal Cangeloso

Printed in the United States of America.

Published by O'Reilly Media, Inc., 1005 Gravenstein Highway North, Sebastopol, CA 95472.

O'Reilly books may be purchased for educational, business, or sales promotional use. Online editions are also available for most titles (*http://my.safaribooksonline.com*). For more information, contact our corporate/institutional sales department: 800-998-9938 or *corporate@oreilly.com*.

Editor: Brian Jepson
Production Editor: Iris Febres
Proofreader: Iris Febres
Cover Designer: Mark Paglietti
Interior Designer: David Futato
Illustrator: Sal Cangeloso

June 2012: First Edition

Revision History for the First Edition:

2012-07-02 First release

See *http://oreilly.com/catalog/errata.csp?isbn=9781449334765* for release details.

ISBN: 978-1-449-33476-5

[LSI]

Contents

Preface

Light-emitting diodes are the future of lighting. Just as the incandescent gave way to the compact fluorescent lamp (CFL) in our homes and offices, the CFL is yielding to the light-emitting diode. These are inherently slow processes given the number of bulbs that need to be replaced and the lifetime of those bulbs, but a tectonic shift is underway. In a few short years, the household incandescent will be a quaint thought, and CFLs will be looked upon as a misguided, poison-laced stepping stone.

LEDs are, of course, nothing new. The technology has been around since the early 1900s, and for years we've seen LEDs in almost all our electronic equipment, regardless of the device, its function, or its maker. For decades they have been affordable to purchase and cheap to operate, but they've largely been relegated to the red, blue, and green status indicators on our computers, radios, and routers. Powered by just a few milliamps and usually outlasting any device they operated within, LEDs served their purpose but were far from fulfilling their potential. Recently, high-power, high-quality LEDs have started lighting our homes and offices, and the big lighting companies —General Electric, Osram Sylvania, and Philips—as well as a number of competitors, are pushing them into the mainstream.

According to the US Department of Energy (DOE), lighting consumes over 14% of all the electricity we use. That means lighting is large industry, but it operates under a number of masters. Not only do organizations like the Department of Energy get involved in lighting, but so does Congress, and all businesses have to take a stance, as does anyone concerned with their energy footprint. Over the course of the next few years, cost-conscious consumers and energy-savvy businesses will ultimately decide if any product or underlying technology is viable or not. And the stakes are not insignificant: according to the DOE, future changes will yield a 19% drop in the energy consumption of lighting. That drop is forecasted to be 46% by 2030 according to a January 2012 report examining the "Energy Savings Potential of Solid-State Lighting" (*http://bit.ly/MPXXHR*).

You wouldn't be here if you didn't already know what a light-emitting diode is, but you should remember this: an LED is just a source of light. How that light is generated—by running electrons through a semiconductor, resulting in a process known as electroluminescence—is fundamentally different than how the incandescent, filament-based bulb works. That our light is being provided by a semiconductor, not the heating of a material (electroluminescence vs. incandescence), is the key to everything you'll read in this book... and the future of mankind's light.

We're at the start of a revolution in home, commercial, and public lighting that will be the biggest shift in the sector since the development of the tungsten filament over 100 years ago. Solid-state lighting is the future and the LED is the engine moving it forward. In a few years the market will almost certainly have settled on LED bulbs (not entirely, but predominantly). Before then, the technology has a long way to go and we as consumers have a lot to learn.

Conventions Used in This Book

The following typographical conventions are used in this book:

Italic
: Indicates new terms, URLs, email addresses, filenames, and file extensions.

`Constant width`
: Used for program listings, as well as within paragraphs to refer to program elements such as variable or function names, databases, data types, environment variables, statements, and keywords.

This icon signifies a tip, suggestion, or general note.

This icon indicates a warning or caution.

Using Code Examples

This book is here to help you get your job done. In general, you may use the code in this book in your programs and documentation. You do not need to contact us for permission unless you're reproducing a significant portion of the code. For example, writing a program that uses several chunks of code from this book does not require permission. Selling or distributing a CD-ROM of examples from O'Reilly books does require permission. Answering a question by citing this book and quoting example code does not require permission. Incorporating a significant amount of example code from this book into your product's documentation does require permission.

We appreciate, but do not require, attribution. An attribution usually includes the title, author, publisher, and ISBN. For example: "*LED Lighting* by Sal Cangeloso (O'Reilly). Copyright 2012 Sal Cangeloso, 978-1-4493-3476-5."

If you feel your use of code examples falls outside fair use or the permission given above, feel free to contact us at *permissions@oreilly.com*.

Safari® Books Online

Safari Books Online is an on-demand digital library that delivers expert content in both book and video form from the world's leading authors in technology and business.

Technology professionals, software developers, web designers, and business and creative professionals use Safari Books Online as their primary resource for research, problem solving, learning, and certification training.

Safari Books Online offers a range of product mixes and pricing programs for organizations, government agencies, and individuals. Subscribers have access to thousands of books, training videos, and prepublication manuscripts in one fully searchable database from publishers like O'Reilly Media, Prentice Hall Professional, Addison-Wesley Professional, Microsoft Press, Sams, Que, Peachpit Press, Focal Press, Cisco Press, John Wiley & Sons, Syngress, Morgan Kaufmann, IBM Redbooks, Packt, Adobe Press, FT Press, Apress, Manning, New Riders, McGraw-Hill, Jones & Bartlett, Course Technology, and dozens more. For more information about Safari Books Online, please visit us online.

How to Contact Us

Please address comments and questions concerning this book to the publisher:

O'Reilly Media, Inc.
1005 Gravenstein Highway North
Sebastopol, CA 95472
800-998-9938 (in the United States or Canada)
707-829-0515 (international or local)
707-829-0104 (fax)

We have a web page for this book, where we list errata, examples, and any additional information. You can access this page at:

http://oreil.ly/SC_LED

To comment or ask technical questions about this book, send email to *bookquestions@oreilly.com*.

For more information about our books, courses, conferences, and news, see our website at *http://www.oreilly.com*.

Find us on Facebook: *http://facebook.com/oreilly*

Follow us on Twitter: *http://twitter.com/oreillymedia*

Watch us on YouTube: *http://www.youtube.com/oreillymedia*

Acknowledgments

Perhaps my favorite part of writing this book was the great people that met along the way. I'd like to extend special thanks to David Cardinal, Julian Carey, John Chu, Tom Riordan, Brett Sharenow, Bert Tao, Deanna Siste, Eric Holland, Andy Beck, Tom Dalton, Brian Wilcox, Susan Larson, and my technical editor, Richard Stevenson. This book would not have been possible without the help (and patience) of Brian Jepson and Gena Vacanti.

1/Opening Remarks

First off, it's all about the lumens. Lumens are the visible light given off by a source, and they are the ultimate goal. Wattage has often been confused with lumens, thanks to how incandescent bulbs are rated (everyone knows that a 60W bulb is not as bright as a 100W one), but more efficient lighting technologies have greatly changed the relationship between wattage (essentially power usage) and lumens (light output). For example, an incandescent might operate at about 12-15 lumens-per-watt (lm/W), while an LED bulb will be in the range of 40-50, a much greater luminous efficacy (Table 1-1). CFLs stack up well when it comes to lumens-per-watt, but they don't have the lifespan of LEDs. CFLs have all sorts of other issues that we'll address later, such as problems with disposal. And those LED bulbs? They are just today's basic, commercially available LEDs—cutting edge models can be much more efficient. These models might produce over 150 lm/W, and even higher than that in laboratory settings.

Table 1-1. *Wattage equivalency and lumens (from Energystar.gov (http://1.usa.gov/76UW2D))*

Wattage (W)	Lumens (lm)
25	250
45	450
60	800
75	1110
100	1600
125	2000
150	2600

The incandescent bulb is a good place to start with any talk about lighting. This design has had tremendous longevity (over 130 years) and it makes for a cheap, versatile bulb. Unfortunately, this design is also power-hungry, inefficient, short-lived (with some exceptions) (*http://bit.ly/4ywsuv*), and fragile. They produce a minimum amount of lumens-per-watt, though they've made appreciable gains over the years, and are highly sensitive to power conditions. For example, a 5% reduction in voltage could double the life of a bulb while only decreasing light output by 20% (*http://bit.ly/LnY7EP*).

One of the most notable strengths of the incandescent is the quality of the light it provides. This isn't as easy to define as some of the other characteristics that will be covered, but it's an important one when it comes to consumer adoption. After all, it's nice to try to sell people on longevity and power savings, but if they think that the new bulbs are ugly or are too different from what they know, you'll find them hoarding 75W and 100W incandescents before such bulbs are removed from the shelves.

Quality of light means that in order for people to be comfortable with the light these bulbs provide, the bulbs will need a color temperature that they find to be in an acceptable range, a high degree of color accuracy (usually measured by CRI), and a usable light pattern, to name a few qualities. The bigger point, as any early CFL or LED bulb buyer could tell you, is that if the bulbs don't produce attractive light that people are comfortable being around, it won't matter how long they last or how little power they consume.

Incandescents have good qualities, but ultimately their inefficiency means they are not a viable solution moving forward. Even modern incandescents can turn about 90% of the energy they take in into heat, which is obviously wasteful and inefficient in the extreme. Physicists might argue that this isn't wasteful at all, and you might enjoy the heat they provide, but most of us want to leave the lighting to the lights and the heating to our furnaces. Before we demonize the long-standing bulb design, it's worth noting that there is such a thing as efficient incandescence. While these are in fact more efficient versions of the incandescent bulb, they are still not at the level of top CFLs and LEDs. In fact, GE was working on a high-efficiency incandescent (HEI) for about 18 months, but gave up on it in order to focus its efforts on LED and organic LED (OLED) bulbs (*http://bit.ly/LXsoew*). HEIs were said to produce about 30 lm/W with the ultimate goal of doubling that amount (*http://bit.ly/LDMI1Y*).The halogen lamp is a type of incandescent that operates hotter and lasts longer, but its efficiency gains are minimal.

The much-maligned CFL solves some of the efficiency problems of incandescent bulbs, usually producing around 50 lm/W. Unfortunately, each bulb contains a small amount of mercury (about 4 milligrams per bulb), so disposal can be a problem, especially if the thin, usually helical, glass breaks. The bulbs have reasonably long lives, usually rated for 5,000 to 15,000 hours —but they don't last nearly that long if they are used in short time spans as rapid cycling is bad for the bulbs. That means a CFL in a bathroom or closet might not last much longer than an incandescent bulb, despite what it says on the package. In fact, a CFL that runs for an average of 15 minutes at a time might last just 40% of its rated lifespan. Alternatively, a CFL that is used continuously from the first time it was turned on might last close to twice its expected lifespan (*http://bit.ly/MImQFo*).

CFLs saw a big jump in marketshare in 2007, capturing around 23% of the market (*http://bit.ly/MPPUuA*), but have been in decline over the last year or so, despite the bulbs being widely available, affordable to purchase, and much cheaper to operate than incandescents. Part of this is due to an increasing number of consumers learning about the CFL's use of mercury, but current economic conditions also indicate that people have simply been looking for a more affordable option. In that respect, incandescents still cannot be beat.

One of the most important characteristics of LED lighting is that they are solid-state. "Solid-state" might be a term we normally associate with computer parts (as in the solid-state drive) but it's not something the casual LED buyer will ever consider. The concept is quite simple: rather than generating light through burning or gas-discharge, LEDs use semiconductors. The is the most fundamental and important distinction that determines why LED lights have their unique characteristics and will be able to have such an impact on the lighting market. As seen in other industries, semiconductors improve at an exponential rate and have a way of taking over wherever they are used. Lighting should prove to be no different.

Of course, LEDs are just one type of solid-state lighting; there are also organic LEDs (OLEDs) and polymer LEDs (PLEDs). Right now, the LED is the main focus of SSL adoption and its future looks quite promising, thanks to the efficiency gains it brings to the market. OLEDs and their carbon-based semiconductors have potential, but high costs mean they won't be a viable option as soon as standard LEDs.

The advent of solid-state lighting doesn't just mean more efficiency. Just as with the introduction of high technology to other parts of our lives—from our phones, to our mail, to our televisions—light is now high-tech. In this case, it's not the tech that makes the difference, it's that this latest step means our lights could soon be gadgets. Today's technology brings with it intelligence and connectivity, which makes way for lights that can be tracked, controlled remotely, and designed to work with other devices. While the humble incandescent was just a conduit for electricity and output both light and heat, a modern-day bulb can be and do much more.

What does this all mean for the LED lamp? Basically, the time is ripe for growth. LED adoption is low at the moment, but not because purchasing one won't pay off. An LED bulb will pay for itself many times over thanks to its energy savings, but the high initial cost is just too much of a hurdle for many businesses and is unpalatable for even more consumers. As prices drop we'll see a dramatic growth, just as CFLs grew when it was clear that they could lead to long-term savings and could, in fact, provide acceptable light for our kitchens and living rooms, not just offices.

Still Unconvinced?

If you are still unconvinced by all the cool technology and other features of SSL bulbs, perhaps this point, courtesy of Tom Riordan, the former CEO of Exclara, will win you over. During our conversation, Riordan noted that LED lamps are subject to stricter regulation than incandescents ever were. Incandescents, after all, have been around for a very long time and are largely the product of a bygone era. He noted that incandescents, if they were invented today, would never be allowed to be sold thanks to today's safety requirements. Think about it: they are extremely fragile, when you drop one glass shatters everywhere, you could electrocute yourself with a broken one, and they contain a white hot piece of metal that is barely protected! And that's just the start. They are also enormously inefficient and give off great amounts of heat that most people could do without.

2/Key Terms

You can't get very far talking about LED lighting without understanding the lingo. Of course, terms like "lumens" and "kilowatt-hour" are important, but a new technology always introduces new concepts to understand.

Color Rendering Index (CRI) and Color Quality Scale (CQS)

CRI is technically, according to the American National Standards Institute, the measure of how similar colors appear under illumination by a test source, compared to under a reference source that has the same correlated color temperature.

In practice, it's often known as the color accuracy, a numerical rating of the color quality produced by a light source. Incandescent bulbs are the benchmark for the test (scoring 100) while a good LED bulb will be rated at 80 or above (this number will almost always be printed on the back of the box). Incandescents—and the sun, for that matter—are the baseline because they emit light at all points along the visible spectrum, while LEDs and CFLs have a spiked, intermittent pattern. Higher CRI value bulbs are available for graphic designers, art galleries, artists, and so forth. But they increase the cost and most consumers won't notice a considerable difference above 80. When using a bulb with a score below 80, the results will vary based on the qualities of the particular bulb, but viewers will typically observe the color of the surrounding objects to be "dingy." Red and blue tones will appear to be off and skin tones can look unhealthy.

CRI is determined by measuring the color of eight samples and then taking the average of those ratings. While that number is representative of overall color accuracy, this testing method is not perfect. By using an average it means that a particular light source might be very strong in some areas and very weak in others. A particular bulb might be very accurate with Dark Greyish Yellow (known as TCS02), yet work quite poorly on Light Violet (TCS07). This could leave the bulb with a relatively high CRI rating, despite its low-quality light under particular circumstances. For this reason and others—including the fact that the test is 40 years old—CRI is seen as a less-than-perfect tool, but it remains the only one on the Lighting Facts label on bulbs.

There are other, more saturated colors that can be used in order to augment the testing, but these are generally not included in the normal CRI rating. The most notable of these is Strong Red (known as TCS09 or R9), which serves as a stand-in for certain skin tones, food types, and other colors we often encounter. When speaking about color accuracy, lighting companies will often relay their CRI score and then call out a high R9 CRI value if their bulb is capable of it.

When it comes to measuring the quality of light, the CRI is the test used most often. However, it's not the only assessment available. Color Quality Scale (CQS) is a newer testing method under development by the National Institute of Standards and Technology (NIST) and has a number of proponents. Its methods are not unlike those of CRI, but it's a newer model, developed over the last few years specifically with LED lighting in mind. The CQS uses 15 color samples, many of which are more saturated than the eight CRI shades, also operates on a 100 point scale, and addresses many of the problems with CRI testing. For example, CRI treats all deviations from the correct value as being the same, when it's generally accepted that humans prefer overly saturated color to undersaturation, thus making shifts in this direction arguably preferable. CQS factors this in, thus focusing somewhat less on color fidelity and more on perception.

You're probably asking yourself, why does this matter so much? It's not just that great light is important. A bad test is easy to game (or is it that easy gaming makes for a bad test?). This means that manufacturers could design around the test, making for a cheap lamp that does well in testing but is not able to deliver the light quality that buyers expect. This would be bad for the buyers and bad for the industry.

It's also been noted that, as in other industries, outdated testing procedures can hold back progress, as manufacturers are forced to comply with old standards as opposed to advancing at the rate of technology's progress.

Color Temperature

The shade of white light is measured on a temperature scale using degrees Kelvin (K). This is important because we expect to see a certain shade of color from indoor lighting, about 2500K–3000K. Warmer colors are more reddish and have a lower color temperature, while cooler colors are bluish and have higher Kelvin ratings. This, as you've surely noted, is the opposite of what you might expect, with cooler colors actually having a high Kelvin count.

Older LEDs and CFLs typically operated at "cooler," bluer temperatures, and consumers didn't like the hospital-like lighting they provided. New LED bulbs have been able to lower their color temperature to approach that of incandescents, primarily by using improved phosphors. A phosphor, which will be covered more extensively later, is a material that radiates light without having to be heated. They are often used in LED lamps, in tandem with the LEDs, in order to dial in characteristics like the color temperature.

As far as LED bulbs go, a more accurate term than color temperature is the correlated color temperature (CCT). Incandescents have a "true" color temperature, but color temperature is approximated for LED bulbs so their value is understood as being correlated (that is to say that the shades of white LEDs produce don't land perfectly on the Planckian locus like incandescents, but they are close enough that the color temperature scale still works (*http://spie.org/x32415.xml*)). The CCT is sometimes defined as "the absolute temperature of a blackbody radiator whose chromaticity resembles that of the light source," but for our purposes we'll be happy with "color temperature."

The general rule is that warmer tones will result in a less efficient lamp. This is because more red phosphor is needed to achieve the lower CCT and this is less efficient than yellow phosphor. This is the case not because of the phosphor, but because the energy difference between the blue photons emitted by the LEDs and the resulting red photons is greater than between blue and yellow.

Gaussian Curve/Distribution

You can't talk for more than 10 minutes with any LED industry insider without the word "gaussian" coming up in conversation. A gaussian curve is a normal distribution, or what most of us know as a bell curve. So how does this relate to lighting? It's all about spectrum. White light, as any elementary school child could tell you, is made up of a combination of all the colors. It is not necessarily made up of the same amount of each shade, or wavelength, though. The spectral power distribution of many LED lamps tends to be gaussian, or normally distributed, with peaks at certain points. Those peaks will vary based on the design of the particular bulb—some will be better with red tones while others might be optimal for green or blue.

Haitz's Law

Computer geeks may think of Haitz's Law as basically Moore's Law for LED lighting. The law states that every 10 years the cost-per-lumen from light sources will drop by 10x while the light produced from an LED will increase by 20x. Put more simply, LEDs will get better and cheaper over time.

The contrast to Moore's Law is that LEDs only need to reach a certain point when they are cheap enough and bright enough for the necessary applications, while we will always have a demand for exponential growth in computing. This is why we know LED bulbs will replace CFLs: in just a matter of years they won't cost $30, they will cost $15, or $10, or possibly $5. The LEDs sometimes account for up to 60% of the manufacturer's costs (though this varies based on the bulb in question and who you ask), so reducing their price is an important part of increasing sales.

The key is that this is a steady trend, one that's been observed since the 1970s. It's not a scientific law in the strict sense, but it means that we can likely count on LED prices dropping in the future at a predictable rate, and we can plan accordingly.

Lamp Versus Bulb

These terms are essentially interchangeable. A lamp is simply something that produces light, just like a bulb. However, not all things that produce light are what we would describe as "bulbous," so the term "lamp" is the catch-all.

The typical incandescent bulb is known as the A19. It carries this name because it has an "A" shape and is 19 multiplied by 1/8-inches in diameter (2 3/8-inches) at its wide point. So your normal, everyday bulb will be an A19 with an Edison screw base. And any trip to the hardware store will tell you that's not the only type of bulb—you might have a light fixture that uses the GU (bayonet-style) base and a parabolic aluminized reflector light (PAR) shape, or any number of other options, but the A19 is the most popular.

Within the industry, the term "lamp" is popular because not all lights are bulbs. In fact, when it comes to solid-state lighting, lots of the lights come in non-bulb varieties, such as overhead LED arrays like Lighting Science Group's Flat LowBay. In the future, an increasing number of lights will be fixtures, which means they will be moving away from the traditional bulb shape because it's no longer necessary.

LM-79, LM-80, and Other Tests

Lighting is a highly regulated field. Agencies like the Department of Energy and companies like United Laboratories track products closely and make sure they adhere to certain standards, so that bulbs don't suck down more power than they say they will or cause pesky fires. LM-79-08 is a testing procedure for solid-state lighting that looks at criteria like efficiency (lumens-per-watt), power usage, total flux, and chromacity. LM-80 is the test for the maintenance, or longevity, of a solid-state light source. It tests bulbs for thousands of hours (6,000 at the minimum) over a range of different temperatures. It only offers recorded test data; no data is extrapolated based on the observed results.

The useful lifetime of LED bulbs cannot be fully tested, not just because of their extremely long lives but because by the time a thorough test was even close to being finished, technology would have passed that model over and replaced it with another. As such, tests are designed to work around this issue. A valid sample size is determined and those bulbs are tested for a set amount of time. After a thousand, or so, hour "seasoning period"—where the bulbs can actually brighten—bulbs are tested for another 5,000 hours. During this time, light levels and bulb failures are recorded. The resulting chart can be extrapolated to determine the point at which the life of the bulb will be over. This is typically at the time when they produce a total of 70% (known as L70) or 50% (L50) of their initial lumens (the group behind the test says either "lumen maintenance life" an be used). The life is determined by the Illuminating Engineering Society of North America's (IESNA) LM-80-08 testing procedure.

When rating the lifetime of a lamp, certain conditions are maintained. There is a set ambient temperature (25° Celsius), bulb running temperature (usually under 90° Celsius), and amperage (generally 350 milliamps) so that conditions are the same from one test to another, and real-world conditions are being reflected.

Tests are devised and accepted by groups like the Illuminating Engineering Society of North America. Hence the full name of a test might be: "IESNA LM-79-07, Illuminating Engineering Society of North America, Approved Method for the Electrical and Photometric Measurements of Solid-State Lighting Products."

Watt Equivalency

Incandescent bulbs were rated by their wattage consumption number. This worked because they were the only lighting option, so we could tell that a 60W bulb was brighter than a 40W one. This (admittedly poor) rating system doesn't work for LED bulbs because they can produce the same amount of lumens using less wattage. So, in order to sell bulbs to consumers who don't know the rating system, LED bulbs are often sold by their wattage equivalency. For example, a Samsung A19 LED bulb runs at 10W but produces 550 lumens. That lumen number means it has a 40W equivalency.

3/LED Basics

LEDs come in a number of shapes and forms (including the organic LED and polymer LED), but the basic LED is a semiconductor. They are small, long-lasting, and power efficient. They are made of a semiconductor material in order to produce the desired color tones (*http://bit.ly/OXk6qz*).

How LED works is not particularly complex from a scientific point of view, but it's quite far removed from the purchase and usage of LED lighting. Lots of information is available if you'd like to dig deeper into the physics and chemistry behind semiconductors, but it should suffice to say that an LED is a diode—a device that allows electricity to pass through it in a single direction —that emits light. That light can range from infrared (non-visible) through the color spectrum, depending on the semiconductor material that is used. For example, a red LED might use aluminium gallium indium phosphide (AlInGaP) and a particular voltage drop (from one side of the diode to the other) to emit the desired color.

White light is the combination of all of the colors in the visible spectrum. Because of this it can result in a number of shades, and producing it is not as easy as just using a particular semiconductor material. There are a number of different methods for producing white light, some of which are more energy efficient than others, and some methods are patented by particular companies. Once you factor in this information and secondary facts, such as the blue LED light is cheaper to produce than white light, then you can start to understand why LED lighting is such a scientific enterprise.

It's worth understanding that there are a few factors that make LEDs an excellent light source. One major strength is lifespan; not only can most LEDs last for tens of thousands of hours, but LEDs don't immediately die. Rather, LEDs experience lumen depreciation, which is a fancy way of saying they lose brightness over time. A well made, properly cooled LED might only lose 5% of its brightness over 20,000 hours of operation. However, this drop off will depend on the cooling and power consumption—LEDs can get more power pushed through them to produce more light (think of this as overclocking it), but it will shorten the life. Heat also has a major role in LED lifespan. LEDs that run in high temperature settings will not last as long as they otherwise would. This is why cooling is so important and many LED bulbs have large, metal heat sinks.

It's interesting to note that while LED bulbs get less bright over time, this lumen depreciation happens in all lighting. In incandescent bulbs, tungsten loss over time means a decrease of about 10%–15% over the 1,000-hour life of the bulb. In CFLs, there can be as much as a 20% loss over a bulb's 10,000-hour lifetime.

Other strengths of the LED include quick start up times, the ability to cycle on and off frequently without damaging the light producing parts, and ruggedness (these are solid-state gadgets, after all).

And the weaknesses? Currently, LEDs are expensive (though that's changing quickly), they are greatly affected by their operating temperature, and, as stated before, they can produce low quality (that is to say, low CRI) or overly blueish light if measures are not taken to correct for this. Finally, LEDs are directional, so manufacturers must ensure that the lamp's housing can direct the light if necessary; while consumer bulbs are often omni-directional, other types, like PAR and MR bulb shapes, are highly directional.

Interestingly, the companies that sell LED bulbs aren't always the companies that manufacture the actual LEDs. LEDs are generally made by companies like CREE, Nichia, Osram Opto Semiconductors, and Philips Lumileds. Manufacturing is important because it can affect pricing, which is the single most important factor in consumer lighting. Companies like Switch Lighting are manufacturer-agnostic, so they can buy LEDs based on who offers the best deal at a given time and whose product fits their specifications and quality requirements.

Despite their extended lives, all LEDs eventually die out. As stated before, LEDs don't stop suddenly, they lose brightness over time. So then, you might be wondering, how companies evaluate their lifespan. LED manufacturers and the US Department of Energy typically use the 70% brightness mark (L70) as the point where the "useful life" of a bulb is spent. The amount of time it takes to get to this point will vary based on the conditions in which a bulb is used, how well it's cooled, and how the parts, particularly the electronics, hold up. After all, just because the LEDs can last a long time doesn't mean that the power supply and other electronics can last that long. This is one reason why lamp design is much more complex than simply providing power to a few tiny circuits.

Example 3-1. DIY LEDs

One of the best ways to learn about LEDs is to put down the books and try them out for yourself. While there is all sorts of interesting science involved and bulbs are quite complex, the basics of LEDs are right out of a high school science class. There are pre-made kits available online that will give you all the parts and instructions you need in order to start tinkering with LEDs, and to understand them at a basic level.

The most basic LED kit you can build is an LED throwie. It's nothing more than an LED and a power source—in this case, a battery (usually a CR2025 or CR2032). The best about this project is that you only need to know two facts: the sides of the battery and the sides of the LED. Each of the components has just two sides. The LED will have two wires coming out of it: the short one is the cathode (the negative side) and the long one is the anode (the positive). The same is true with the battery, where the flat side is the positive and the smaller, inset side is the negative.

To construct an LED throwie you simply need to match the positives and negatives, creating a circuit. (Usually there is a magnet involved so that it will stick to metal objects, like a refrigerator.) Pretty simple, right?

If you're at all familiar with the basics of electronics—maybe you've put together an kit or two—then you might know that a simple LED circuit usually has a resistor in place. The resistor limits the amount of current that goes to the LED from the power source. By using the correct resistor, you'll ensure the longest possible life for your LED and prevent any possible damage to it. Basically, what it comes down to in this case is that you could use a resistor, but you probably don't have to. The LEDs will operate along a curve where more current means less lifetime, and since LEDs and batteries like these are cheap, they are easily replaceable.

If you want to stick with the DIY angle, and learn more about the circuitry, switches, and resistors try out a kit like those from Maker Shed, LittleBits, or Sparkle Labs. Maker Shed's Mintronics: Survival Pack (*http://bit.ly/MpfcCv*) is a cheap way to get started: it includes a mini breadboard, some LEDs, resistors, and other components such as a 555 timer chip, which can blink LEDs.

If you want to take things to the next level, you could seek out the MinM (*http://bit.ly/NX1VfS*). This little gadget combines a programmable LED with a bit of silicon. It's smarter than the typical LED bulb, but it's ripe for all sorts of experimentation with computer languages, wearable electronics, and soldering.

4/The Bulb

There isn't a whole lot to the typical LED bulb. In addition to the standard gear —the base, lens (optics), and/or reflector—there is a driver, which is explained in more detail below; PCB, LEDs, an encapsulant over the LEDs, a phosphor, and usually some sort of a heat sink. The ideal scenario (or the goal for some manufacturers, at least) is for the LEDs, hardware (cooling materials, etc.), and driver to each make up for one third of the cost of the bulb, but that's essentially just a rule of thumb—real-world demands and technology issues often get in the way of such ideals. As with any other manufactured product, it's all about tradeoffs. By investing more in the cooling, it's possible to run more power through the LEDs, which means you can use fewer of them to get the same amount of light. Similar tradeoffs must be made, for example, between the quality of the light and the efficiency of the bulb.

The driver is essentially the lamp's power supply, but because this is a modern light source, there is some intelligence built in as well. The driver, which usually lives at the base of the bulb, is able to not just transform the energy the bulb takes in to what the LEDs need, but also to do things like throttle the power being sent to the LEDs if it senses that the unit is too hot. This will not only prevent possible damage to the surrounding area but it will extend the life of the LEDs.

The driver is one area where bulb manufacturers—who often use LEDs from the same providers—can compete with one another in terms of size and electrical conversion efficiency. Drivers can be AC or DC, both of which have their advantages, and they can be as complex an exercise in integrated circuits as any electronic product in this price range. For example, AC systems do not use as much power as DC ones, but they are able to transfer less to the LEDs because of the inherent properties of the alternating current. AC-AC power supplies use less parts and are cheaper to produce than AC-DC ones, but often result in a less efficient bulb, when comparing the total amount of power taken in to the lumens it puts out. As with other parts of the lamp, there is a constant push to lower cost, which often means using fewer components.

Because LEDs do not run off of the same level of current that is provided to buildings—120V in the United States—the driver must perform a power conversion. Many LEDs are low voltage devices (at least, relative to standard line voltage), so a decline must take place. High voltage LEDs do exist, but those in many lamps might need just 3.3V or so. This requires the use of small, dependable electronics that will be able to power the bulb as long as the LEDs inside it will live. Components like electrolytic capacitors and MOSFETs must be used in order to deliver consistent power drive the LEDs. Add in the dimming (possibly through pulse width modulation [PWM]) hardware and an intelligent LED controller and you have a complex piece of electronics that replaces an incandescent...which had no problem using the main's voltage.

One of the conundrums of today's LED bulbs is that because bulbs are often oriented upside-down and in areas with limited air circulation (like a ceiling can), the trusty A19 shape can become a serious problem. Because the bulb is oriented upside-down the driver is at the top, which is where the bulb's heat flows. Special precautions have to be taken when designing a bulb so that the driver does not overheat. If a bulb were to overheat it would initially reduce the lumen output, but in extreme cases it could damage the LEDs as well as the interior circuitry, possibly creating a fire hazard (*http://1.usa.gov/dJ2F0s*).

The heat sink, the metal structure coming up from the base of most bulbs, is one of the most noticeable parts. It is designed to dissipate heat from the LED out to the surrounding area. Thanks to the LED's high efficiency, not much power is turned into heat but the cooler LEDs run the longer their lives are (to a point), so it makes sense to include these. All high-quality, brand name LED bulbs will have some sort of passive cooling as it's an important part of transferring heat away from the sensitivity components.

But why does an LED bulb need a heat sink when none of the other major bulb types require one? The main reason is that LEDs don't give off heat in the form of infrared radiation. This means cooling must be handled through other means, such as conduction through a heat sink. Interestingly, consumer LEDs don't give off ultraviolet (UV) light either. This happens to have a very useful side effect: LED bulbs don't attract insects, which are drawn to UV light.

The heat sink shouldn't be written off as just another part of the bulb: cooling is a critically important part of the design. We know heat limits LED life, but that's not the only reason to keep cool. There is a short term impact to heat as well: hot LEDs produce fewer lumens.

In addition to the LED's sensitivity to heat, the aforementioned drivers are often designed to throttle back light output when too much heat is detected in order to cool the lamp and preserve its life, and to prevent parts from being damaged. But as destructive as heat is on the LEDs, there is another important reason to limit heat build-up: the Underwriters Laboratories (UL) will not approve any lamp that goes above 90° Celsius, as per the UL 8750 safety standard.

LED bulb manufacturer Switch Lighting has taken an extra step with cooling by filling their bulbs with liquid. Their design not only has metal heat sinks on the interior of the bulb but the use of a special liquid promotes cooling through convection as well as conduction. This further dissipates heat, giving

Switch better cooling which, in turn, means the company has been able to design brighter bulbs than some of their competitors, including 75W- and 100W-equivalent bulbs (1180 and 1750 lumens, respectively). Better cooling also means that a manufacturer could theoretically use fewer LEDs and push them harder in order for a bulb to produce the same amount of light, but would have to trade off lifetime for price.

The inability of LED bulb makers to produce high-lumen (essentially higher than 60W equivalents) bulbs has been problematic. It has created the perception among consumers that LED bulbs are unable to produce the equivalent amount of light as a 75W or 100W bulb. This has, so far, limited LED bulb applications to standard task, ceiling, and floor lighting. As designs improve and LEDs become more efficient, higher wattage-equivalent bulbs will become available, though their success as products will be limited by their price. As with anything else, these prices will drop eventually, but it will take time before a 75W incandescent can be affordably replaced by an LED bulb.

One question that is invariably asked when it comes time to buy an LED bulbs is, why are some of them yellow? (Or, more often, why would I buy something that looks like a bug lamp?) This happens with models like the popular Philips AmbientLED, which has an aluminum heat sink crowned by a deep yellow, plastic-y bulge that is split into thirds. The packaging makes it abundantly clear that while the top is yellow, the light it emits is the standard white that you'd expect from any household bulb. But why the yellow and why the bulge?

It turns out that the latter is the easy part to explain. The extension of the light-producing bulb elements means that the light that leaves the bulb is able to reach a larger amount of its surroundings. This is to say that the bulb is less directional than it otherwise would be, and less directional than an older style of LED bulb. If you were to look at the pattern of the light the bulb throws, you would see that it offers better coverage to the area behind the base than a typical LED bulb does (*http://bit.ly/LLecqH*).

And what about that odd, off-putting yellow? That's because of the bulb's phosphor (not phosphorus). A phosphor is a general term for any material that radiates light without having to be heated up to do so. A glow-in-the-dark toy uses one type of phosphor, but so does something as simple as petroleum jelly, which will glow when held under a black light (an ultraviolet light).

How this ultimately works is that the LEDs inside bulbs are often blue (using a semi-conductor like indium-gallium-nitride [InGaN]), not white. When the blue light contacts a phosphor, the result is white light. Some bulbs use LEDs that are coated in a phosphor; in this case, Philips has put the phosphor on external components, creating what is known as a *remote phosphor*. The blue light contacts the phosphor, and the mix of blue and yellow produces white light.

Of course, not just any white light is OK—people want a particular color temperature for their environments. This is usually 2700K–3000K for an American home, but preferences can vary in different countries. In Asia (Japan especially), higher color temperatures are more popular. In other parts of the world, people prefer the nearly amber light of 2500K bulbs. Because the phosphor is the largest single factor when it comes to color temperature and CRI, it's the most important decider when it comes to a critical part of the lighting discussion: light quality.

Light quality is one of the reasons early LED bulbs didn't take off and why CFL adoption was slow. Buyers, especially consumers, have a set of expectations for their lighting purchases, which takes into account brightness, price, temperature, and CRI. Maybe they don't know the terms, and maybe reading the back of the box won't help much, but once the light is home and operating, you can be be sure a lot of people were not happy with that blue-tinted CFL (which probably also took a long time to warm up and didn't work with their dimmer).

Light quality is a characteristic where lamps just need to reach a certain point: essentially, an acceptable color temperature and then 80-90 CRI. After that point, you're off in the land of specialty bulbs, where choosy buyers can have a color temperature they prefer or an artist wants 95+ CRI bulbs to produce top notch color accuracy.

And why not just manufacture all bulbs at a high CRI level? The answer is just what you'd expect: cost. Proper color rendering isn't as easy as just using the right components; it requires research and high quality materials. There is also a documented tradeoff between color rendering and efficiency (in terms of lumens-per-watt). With today's manufacturers almost completely focused on making consumer bulbs cheaper to produce, and factoring in the never-ending push to maximize efficiency, color quality can fall by the wayside.

At some point, everyone interested in solid-state lighting probably wonders why some opt to use a blue LED—known as the emitter—and a remote phosphor. It's all about efficiency, grabbing up to a 30% gain (*http://bit.ly/LLehuH*) according to one source, albeit the CEO of a remote phosphor-focused company. It's best to break down claims and design choices like this into their component parts, the first being, why use a blue LED in the first place? This is a fundamental concept of today's LEDs; blue LEDs are just more efficient. This can be explained with scientific terms like "quantum-confined Stark effect," but the larger point is that blue LEDs are just more efficient, enough so that it makes more sense to use them and a phosphor as opposed to trying to product white light.

And why place a phosphor apart from the LED as opposed to just combining the two? Not all companies use a remote design, so while this not is necessarily better, there can be advantages. One advantage has to do with LED placement. Since LEDs need to be kept as cool as possible for them to run efficiently, by placing them against the body of the bulb, they can use the heat sink to be cooled via conduction. Better cooled LEDs can mean cheaper lights—if fewer LEDs are used and they are pushed harder—or brighter bulbs, if they are simply pushed harder. With the LEDs against the heat sink and the light becoming diffused through the remote phosphor, it's possible to design an omnidirectional bulb any number of different LED placement patterns.

The remote phosphor can control light quality, as well as act as a diffuser for the light, optimizing the pattern and filling a room better than the highly directed LEDs alone. It can also reduce the appearance of bright spots on the bulb, as well as make it easier for manufacturers to manage the purchase of LEDs (the remote phosphor can be changed to get to a desired CCT or CRI without having to be concerned with the LEDs themselves (*http://bit.ly/QH6jCQ*)).

Currently, remote phosphor is a very good option many manufacturers use. Some of these bulbs, like those from Philips, allow the user to see the remote phosphor—that dark yellow plastic at the top of the bulb that immediately turns off some buyers. Luckily for companies that have opted to go the remote phosphor route, this isn't as much of a problem as it might seem. First of all, it's entirely possible to cover up the phosphor with a another diffuser. This will limit the brightness of the bulb to some degree (it's akin to going from a clear bulb to a frosted one) but the gains of a high quality remote phosphor are often enough to outweigh this. For an example of this, look for one of the new CFLs on the market; there is a good chance that the helical glass tube will be placed inside a frosted bulb. Another point, raised by a representative of Intematix, is that the phosphor doesn't have to be yellow, it can be a lighter shade, even khaki. These muted color options might not make for the best possible efficiency, but if the buyer doesn't need a top CRI score or a perfectly dialed-in color temperature, there is some room for flexibility.

One interesting point with the phosphor is that today's materials completely lack an important characteristic: persistence. So when you turn off an LED it stops producing light immediately (about 20 nanoseconds (*http://bit.ly/LLekqc*)). As noted before, LEDs are highly sensitive to power inputs, and they are often designed to operate at 120Hz, basically flickering on and off faster than our eyes can recognize. How does this all tie together? This explains why LED lamps need good drivers and are sensitive to their power conditions. Even the briefest lapse in power will cause a flicker (they can react in mere nanoseconds) while on an incandescent, the lapse wouldn't matter because the filament will retain heat for that brief period of time (it has persistence). If one day phosphor with a high level of persistence were developed, it could be a big step forward, possibly enabling better lights and cheaper drivers.

The technology is advancing each year, so gains will surely be made on this front, with great strides having already taken place in the last few years. A final factor to keep in mind is that the most important buyer for early LED bulbs—businesses—usually don't care what color the exterior of the bulbs happens to be, as bulbs are often enclosed and, more importantly, because they look totally "normal" when running. While the average consumer is said to operate their lights for three hours a day, the average business runs them for ten. This means that there will probably be no point when a customer walks in and sees a strange yellow bulb, because when customers are present, the light will be operating and they won't appear out of the ordinary.

The alternative to using a blue emitter plus a remote phosphor would simply be to use an LED that produces white light. These LEDs sometimes combine colors (like RGB, for the primary colors red, green, and blue) to create that white light. While many producers are happy to go the blue emitter route, some proclaim the benefits of white mixing colors when it comes to color rendering, control over color temperature, and even efficiency (*http://bit.ly/N4IYYH*). Another option is to use RGBA, the "A" being amber.

What this chapter has covered so far includes merely the basics of LED bulb design. There are many other questions that could be asked about any particular product, including: Does it use high-voltage LEDs? Does it have a few powerful, expensive LEDs, or does it use many cheap ones—sometimes over 100—to produce light? How is the light distributed? Is it directional, or will it give full coverage to a reasonably sized room if placed in the center? There is also the matter of the printed circuit board (PCB), if it has one. That's an important part of many bulbs, as is the specific substrate the LEDs use. These are all important characteristics that, while fascinating, won't affect buyers unless they lead to a particularly important advancement.

One example of such an advancement was developed by a company called Soraa, which is making a significant bet on "GaN on GaN" (gallium nitride on gallium nitride) substrates in their LEDs. The progress comes in the form of the technology's efficiency over other methods like GaN on sapphire, or GaN on silicon carbide. Soraa's research and development has led to its claim that they can use ten times the current as an alternative material. This increase translates to ten times more brightness and what the company believes will be an industry-changing advancement (*http://bit.ly/MIPt6K*).

What's the point to all this? LEDs and SSL can be just was complex as you want to make them, especially if you have a thing for semiconductors or photometry. It's one thing to understand the lighting in terms of heat sinks and LEDs, but it's entirely another to get deeply involved in the electrical engineering, physics, and material sciences that allows these products to improve from one year to the next. Many of these elements are outside of our scope and are fully explained in white papers, journals, and textbooks, if you'd like to dig more.

Before moving forward, there is one more thing we non-doctorates can do to understand how LED bulbs work. Here are a few pictures of the inside of two self-ballasted LED bulbs currently on the market: one from Lighting Science Group and one from Philips. While there is a lot to take away from these

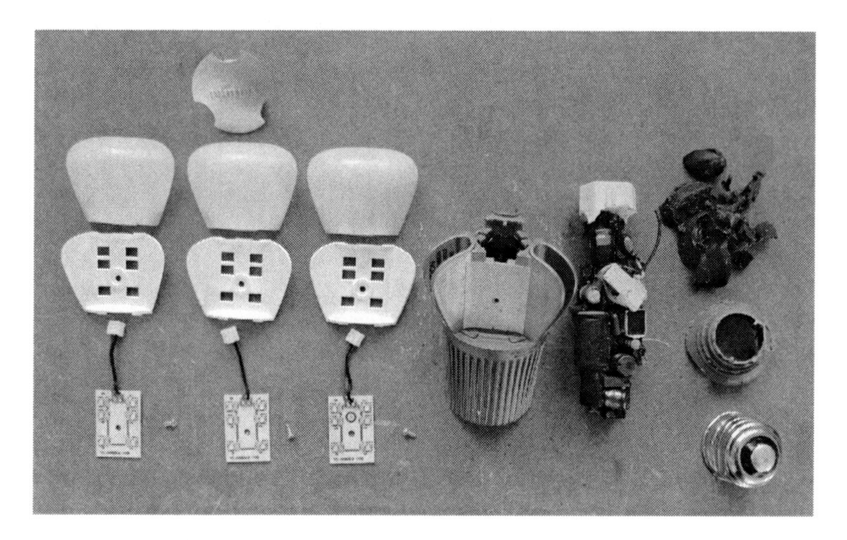

Figure 4-1. *A Philips AmbientLED 12.5W broken down into its components*

photographs, the most surprising fact is there are is no empty space in either bulb. Both are filled with a potting compound that makes them hard to take apart (not that they are user-serviceable) and protects the internal components.

An examination of each bulb gives us some insight into the design philosophy of Philips and LSG as well. Don't miss the high quality connectors and ceramic LED boards, which are backed with a thermal interface material (TIM) in order to best transfer heat to the heat sink. The LSG Definity—A19 Omni V2—Dimmable bulb (Figure 4-4) isn't as nicely constructed as the Philips 409904 Dimmable AmbientLED 12.5-Watt A19 (Figure 4-1), its materials don't feel as nice, the bulb isn't as atheistically pleasing, and the top pieces don't fit together and come apart as the Philips does. (See Figure 4-2, Figure 4-3, and Figure 4-5 for photos of these deconstructed bulbs.) The LSG model has fewer components, which is an important factor, but it actually costs more then the Philips: $27.85 versus $24.34 (at the time of this writing). Both bulbs produce just about 800 lumens, according to their packaging.

Figure 4-2. *The bulb's ceramic PCB*

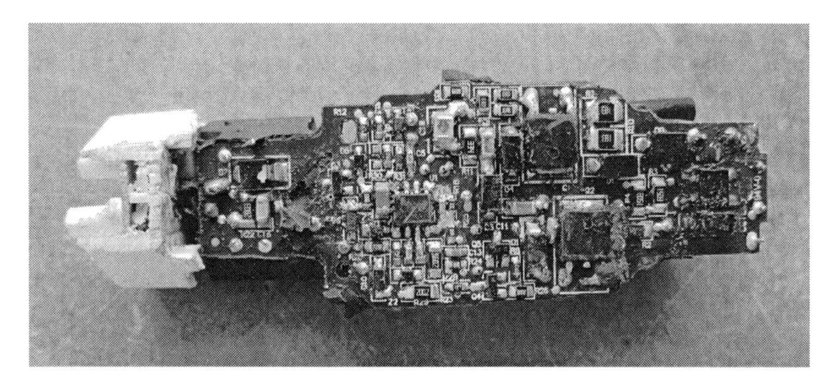

Figure 4-3. *The back of the bulb's driver*

Figure 4-4. *Lighting Science Group Definity A19 bulb—LEDs*

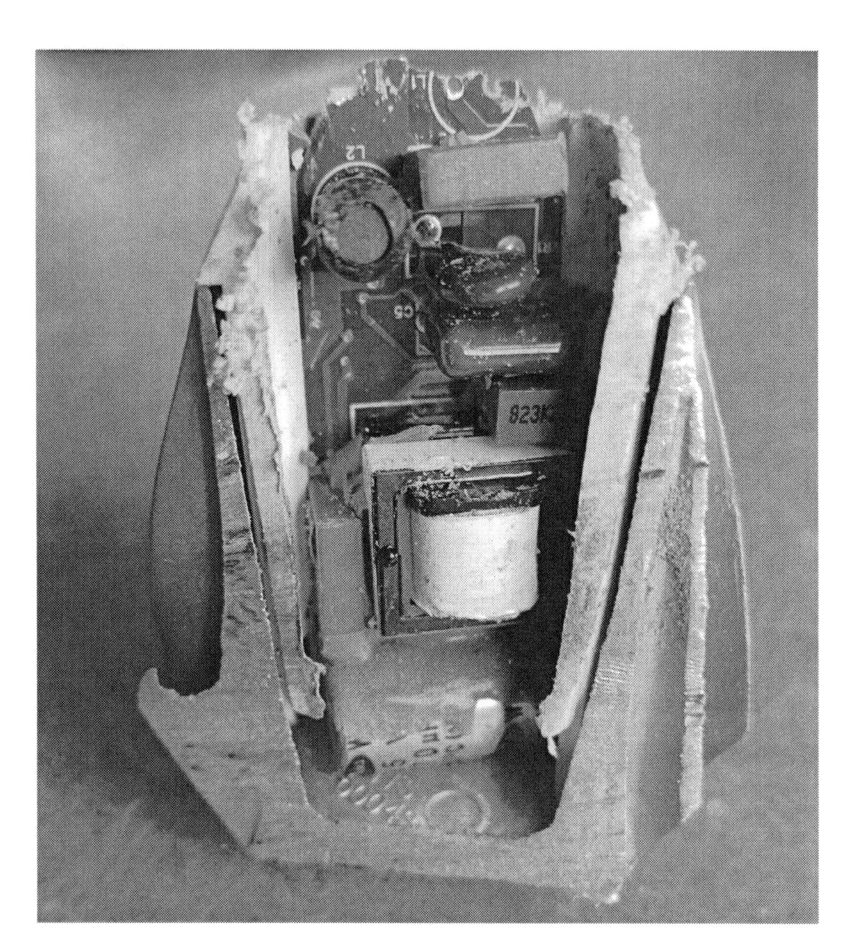

Figure 4-5. *Definity A19 bulb—heat sink and driver*

5/Non-Bulb Lighting

LED bulbs will have so much impact due to the fact that they are direct replacements for the lights people have used for years. A standard, household bulb—the A19—can be easily replaced with an LED model with nothing more than a few turns of the wrist. This means that, using existing infrastructure, tremendous amount of energy can be saved every year...and billions of dollars worth of bulbs can be sold. Just how much energy are we talking about? A 2011 *Energy.gov* report put the potential savings at 84.1 tera-watt hours (TWh) a year if only the A-type bulbs in the US were switched to LEDs (*http://bit.ly/ezmYSu*). And while the standard household bulb is important, it's far from the only application of solid-state lighting. Aside from the A19 there are also PAR, MR, GU, and other styles of bulbs. And those replacement models are just the start. The LED will extend to fixtures, like flood lights, tube lights, bay lighting, and even street lights (see Figure 5-1). There are also LED lamps, in the common use of the word "lamp," like the one found on your desk or side table.

LED lamps have been around for years now. This is partly due to the fact that they don't have the same expectations set for them as their bulb-based counterparts. In the case of a task lamp, LEDs have clear benefits: they don't wear out, they don't put out much heat, and individual LEDs can be placed in series, allowing for clever designs that an incandescent could never pull off. Cost sensitivity with a stand-alone lamp is nothing like it is with bulbs, so a lamp can cost hundreds of dollars and people will buy it, especially if it has the right designer's name on it.

A prime example of this is the Herman Miller Leaf lamp. A creation of super-designer Yves Behar, the Leaf uses a grid of LEDs mounted to a thin metal arm, offering a design and light pattern that would not be possible by using a different light source. The lamp was introduced in 2006, making it a relatively early entrant into the market. At $500, it wasn't (and still isn't) cheap, but it was able to capitalize on its attractive design and eco-friendliness, making it a notable product and Herman Miller's first step into the lighting market; however, people were quick to point out that it wasn't actually very efficient. Its early-model LEDs didn't consume much power, but they didn't produce much light either, so in terms of lumens-per-watt, the lamp is far behind today's technology. The 20 LEDs are said to consume 8 to 9 watts (*http://bit.ly/MOE5mP*), though its lumen production number is unavailable.

We've yet to see a major shift over to LED desk lamps, but this is an ideal application of the technology. There isn't as much focus or scrutiny on these products, so little information is available. Also, prices haven't dropped in the way they have with LED bulbs, but that's due to the size of the market and the fact that the actual LEDs are a small part of the total production cost, not the bulk of it, as it is for a bulb. That noted, there is a full range of LED lamps available today, from no-name task lamps, to premium products from respected companies like Humanscale, Steelcase, and of course, Herman Miller.

Established companies aren't the only ones getting in on the LED action either. Designers are entering the market with cool designs that capitalize on the shapes that LEDs make possible, and grabbing headlines with the LEDs' "37-year lifespan" (*http://bit.ly/qzFZUk*). Some of the most impressive include the CSYS by Jake Dyson, as well as offerings from Artemide and Pablo Designs.

These products have progressed over time, just like all LED lighting. Early products, like Koncept's Z-Bar, were expensive, not dimmable, and featured unattractive blue-ish light. Today, lamp manufacturers are able to benefit from the improved color temperatures and increased user control over brightness levels, making for a much better lamp than the first generation model that was released in 2005.

Task lamps are just one part of the non-bulb market, and are not nearly as important as another category: fixtures. These are generally aimed at businesses and new construction, and they stand to have a large and early impact. That's because businesses are much more sensitive to power costs and they are often more able to withstand a high initial investment, especially when they run the numbers and find that the return in investment (ROI) on efficient lighting is often just a matter of months. Department stores, warehouses, and parking lots, are just a few of the businesses that stand to benefit from switching to LEDs, from options like mercury vapor and high pressure sodium.

The larger point is that LEDs will be a predominant light source in the future, and the bulb is far from the only form that they will come in. Just like LED's can be adopted to fit into a conventional bulb shape, they can find their way into high-design lamps that have very little to do with power-savings, cost-sensitivity, or the other aspects we normally associate with bulb lighting. Aside from lamps, LEDs also provide light for automobile headlights, signage, and LCD backlighting.

Figure 5-1. *Philips Lumec street light in New York City, which uses 80 1W LEDs*

6/Buying an LED Bulb

If you are in the market for an LED bulb (and, for argument's sake, let's keep the medium screw base bulb separate from LED lighting overall, such as lamps with built-in, non-replaceable LEDs) then there are a number of factors to consider. We've covered some of them, and many will be obvious to any consumer: price, appearance, brightness level, power consumption, whether they are dimmable or not, and so on. These are all points that any buyer would consider before purchasing a bulb, regardless of the underlying technology. However, because of the scrutiny applied to solid-state lighting, an examination of the product can be much more in depth than just a casual price/shape/brightness check.

Aside from the aesthetics of any one particular bulb, all buyers should consider something businesses are already acutely aware of: ROI. This is a simple consideration of the fact that while LED bulbs are initially expensive, they are generally cheaper in the long run. This might not have been the case with CFLs—take for example, running in short durations could negatively affect lifespan—but it's generally the case with LEDs because of two factors. These, as you know by now, are their longevity and their efficiency. What buyers also need to consider, though, is replacement cost. If you have a light that is 25 feet high and you need to rent a scissor lift or hire a handyman with a extendable ladder to change the bulb, then the fact that a bulb should last 20 years means it's an even better value compared to one that won't.

One of the great things about lighting, at least if you are a buyer, is that safety-conscious governments and trade organizations take it very seriously. Here in the United States, agencies like the Department of Energy and the Federal Trade Commission (FTC) make sure that a bulb's packaging is covered with all the crucial, test-certified information that consumers need to make their decision. The manufacturers, being the good capitalists and business people we know them to be, will also strive to make sure consumers understand all the benefits and savings that go along with expensive new LED bulbs.

Let's walk through a typical packaging to see what it can reveal.

In the case of the Philips 409904 Dimmable AmbientLED (Figure 6-1) 12.5-Watt A19 Light Bulb (medium base, A-shape) the first thing that pops out is the color of the bulb—it's yellow! As previously explained, that's just the remote phosphor, but the average consumer doesn't know that so Philips makes it abundantly clear that the AmbientLED is white when active. Not only is "white light when lit" written directly above the yellow bits (Figure 6-2),

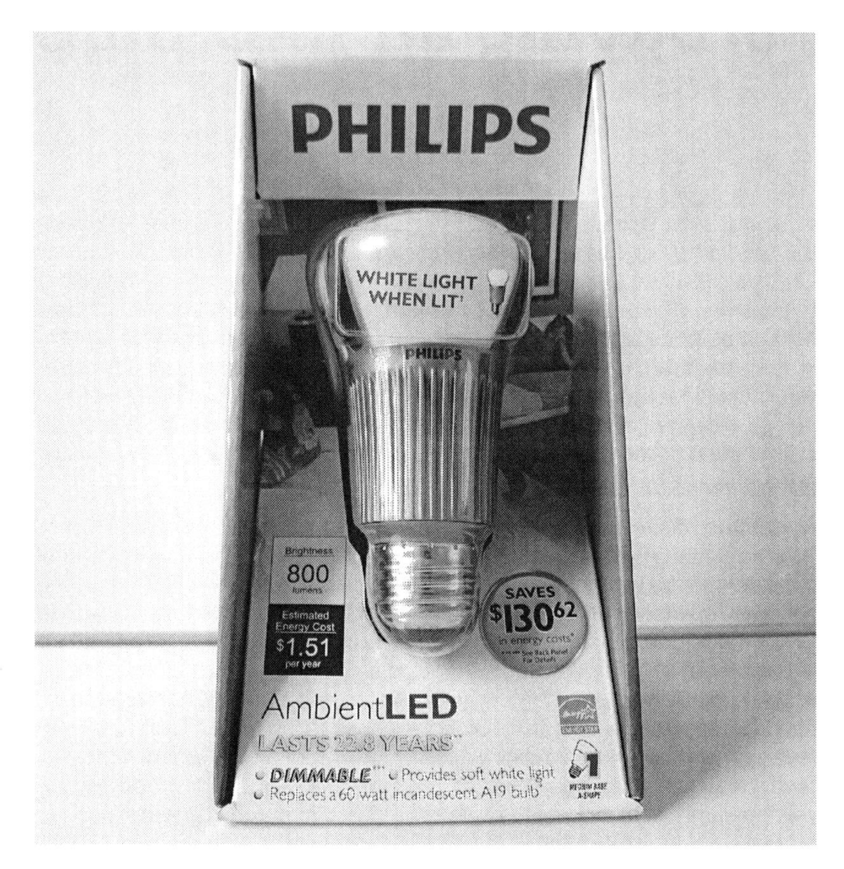

Figure 6-1. *The Philips AmbientLED—front*

but the back of the package displays a yellow lamp that is off and a white one that is on. The company fully understands how dismaying the color can be to buyers, and short of selling a bulb that is running off a battery, this is the best they can do.

After that, the premium space on the packaging is directed at clearing the bulb's next major hurdle: its price. At $26.90 (and an MSRP of $49.99), the AmbientLED isn't egregious, but it will cost you about 20 times more than a standard incandescent bulb ($26.90 vs. $1.32 for a GE A19 60W bulb). Customers understand that, and it's Philips' job to stop people from putting down their product as soon as they spot the price tag. Three pieces of information on the front alone convey this. The first screams, "Saves $130.62 in energy costs" over the 25,000-hour life of the bulb. The second notes that the bulb

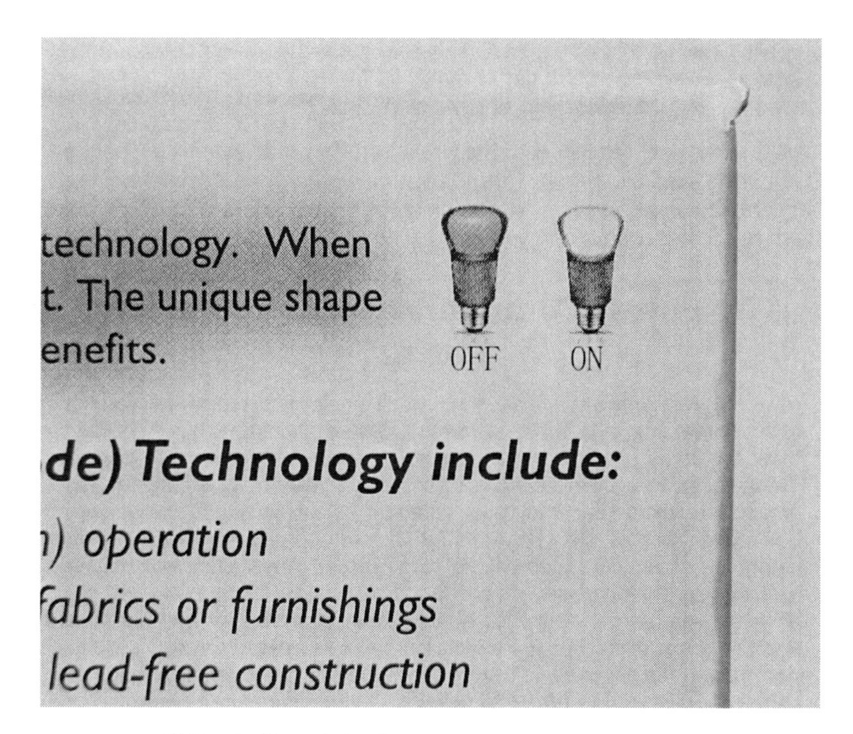

Figure 6-2. *Yellow bulb explained*

costs just $1.51 a year to operate, and the last is that it will work for 22.8 years (when used for 3 hours a day). These won't prevent the bulbs from being expensive, but at least they make it clear that LEDs will be cheaper in the long run.

The packaging also advertises that this LED bulb is dimmable. Not all CFLs can claim that, nor can all LED products. The ability to dim a bulb using the existing controls—which are almost always made for dimming incandescents—is important to many buyers. What's not explained is just how much the bulb can be dimmed. Will it simply turn off when it hits the 40% mark? Will it have increasing lumen levels from 0% straight through to full power? For an LED bulb to do this it requires a dimmable LED driver, which costs the manufacturer extra money—and they want you to know about it.

But guess what? This still doesn't ensure the switch you have at home will work. Most dimmers were built to work with high wattage incandescents, so when you move over to a bulb using a fraction of the power, you'll still be providing the LED bulb with more than enough for it to operate at 100% brightness. CFL/LED specific dimmer switches are available...but your home probably doesn't have them unless its new or you've upgraded. Because it's both difficult and expensive for manufacturers to design their bulbs to compensate for the cheap, poorly designed dimmers in most homes, they probably will never do this very well. A likely solution is that power companies will subsidize, or even give out, better dimmers for their clients, just as they did when they encountered the problems with CFLs.

In case you're not familiar with the subsidies, they did, and still do, happen. Companies like Duke Energy (*http://www.duke-energy.com/freecfls/*) regularly give out free bulbs and rebates on bulbs to customers in an effort to get them to consume less energy.

Why Power Companies Want You to Save Power

Have you ever gone into a store and had an employee expressly tell you not to buy something, or to buy less of what you were planning on buying? Didn't think so. Oddly enough, that happens every day with power companies. These giants, who are in the business of selling us electricity, regularly tell us how to save electricity. They even go so far as subsidizing high-efficiency bulbs for our homes. What's the deal? It's actually surprisingly simple: power plants are extremely expensive. So with current plants aging and usually running near their capacity it's cheaper for the power company to sell less power to you (and a few million other people) than it is for them to build another power plant. There are more complex aspects to this too, such as government regulations and "decoupling" the sale of electricity and the income made by a utility (*http://bit.ly/MxYLDI*), but ultimately it's possible for everyone to win when you save energy.

Finally, the front of the packaging also displays the number of lumens the bulb produces (800 in the case of the AmbientLED, Figure 6-3). This is an important part of the move away from wattage as an indicator for brightness and to inform consumers about what kind of performance they should expect.

There is a lot of information on the back of the packaging, including the features, footnotes explaining the claims on the front, warranty, and so forth, but the most important part is the Lighting Facts section. Note its similarity to the Nutrition Facts label that is on any packaged food product you buy in a supermarket. The design clearly states that this is something that is tested, certified, and as free (as possible) of marketing and advertising. This is the part that the Department of Energy cares about and this is the information you can count on when attempting to make an informed decision. In fact, even the size requirements (*http://1.usa.gov/MDP7PH*) for the Lighting Facts label (*http://1.usa.gov/MPSroM*) match those of the FDA's Nutrition Facts.

The Lighting Facts starts off with the brightness in lumens and then moves on to the estimated yearly energy costs. In this case, the Philips bulb comes

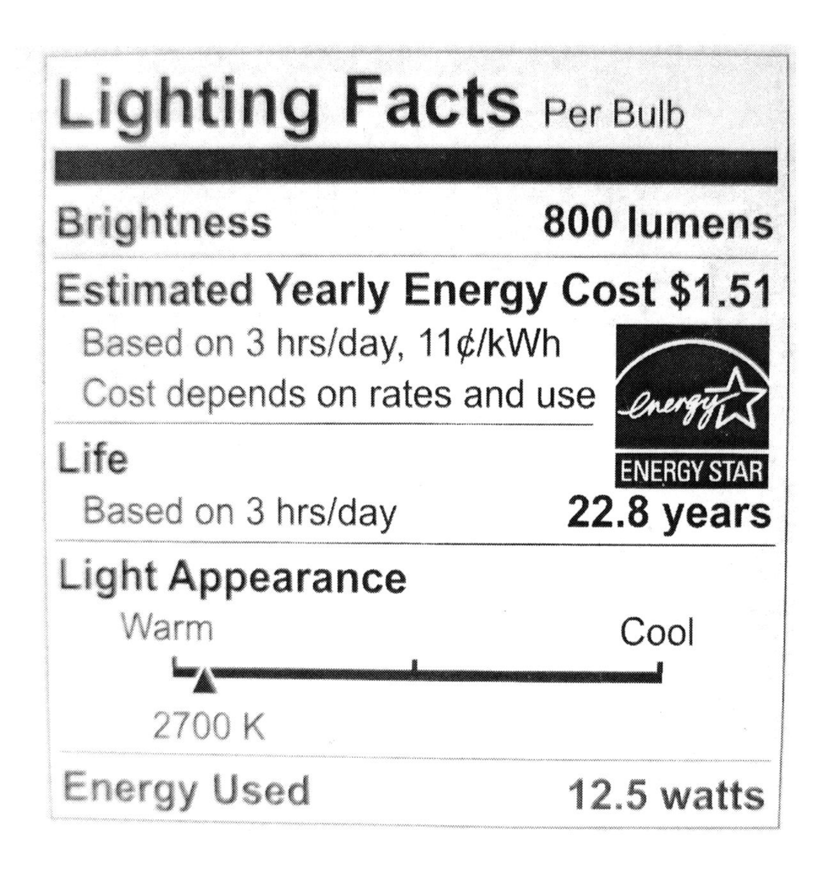

Figure 6-3. *Lighting Facts label—Philips AmbientLED 12.5W—old design*

in at $1.51, assuming 3 hours a day at 11 cents per kWh. This number will vary based on where the bulb is operated, what time you are running it, and who provides your power. Next up is the lifespan of the bulb (22.8 years), again based on 3 hours of usage a day. After that comes the "Light Appearance," which plots the bulb's color temperature on a spectrum that runs from warm to cool. At 2700K, the indicator on this gauge is about 5% from the leftmost point, meaning the bulb is just about as warm as it can be. The final piece of information in the Facts box is the bulb's power consumption, which in this case is 12.5W.

The Lighting Facts label is required on the side or rear of every lighting product, thanks to the FTC's 2010 Appliance Labeling Rule. The black and white box on the front that indicates the brightness and estimated yearly energy cost is actually part of the government-mandated label. The label was officially required for use on all lamps sold after January 1, 2012 (*http://bit.ly/MpgVrp*).

This is actually a simplified version. On the Department of Energy's LightingFacts.com (*http://bit.ly/lightingfacts*), they have an example of an LED Lighting Facts label that has all the above information, plus CRI, CCT, efficacy (lumens per watt), registration/model number, and a more detailed color temperature gauge (that's no longer in black and white). The simplified version has been the official one, in keeping with the look and feel of the FDA's label that we all know so well, but it's really just a start. Right now, the DOE mandates that all Lighting Fact labels must include brightness, energy cost, the bulb's life expectancy, light appearance (for example, warm or cool), energy usage, and whether the bulb contains mercury, on the back of the box. The front must include brightness in lumens and estimated annual energy cost. The rules also state that the lumen output and presence of mercury must be printed directly on the bulb. After March 2012, manufacturers will have the option to also put lumen maintenance (the projected loss of light over time) (Figure 6-4) and warranty information on the label (*http://bit.ly/NeQlbo*).

One thing that's not mentioned in the Lighting Facts, but would be helpful to appear on the box (or at least on the manufacturer's website), is if the bulb is designed to work in an enclosure. Not all LED bulbs are, so you should know what you are getting into before you make your purchase.

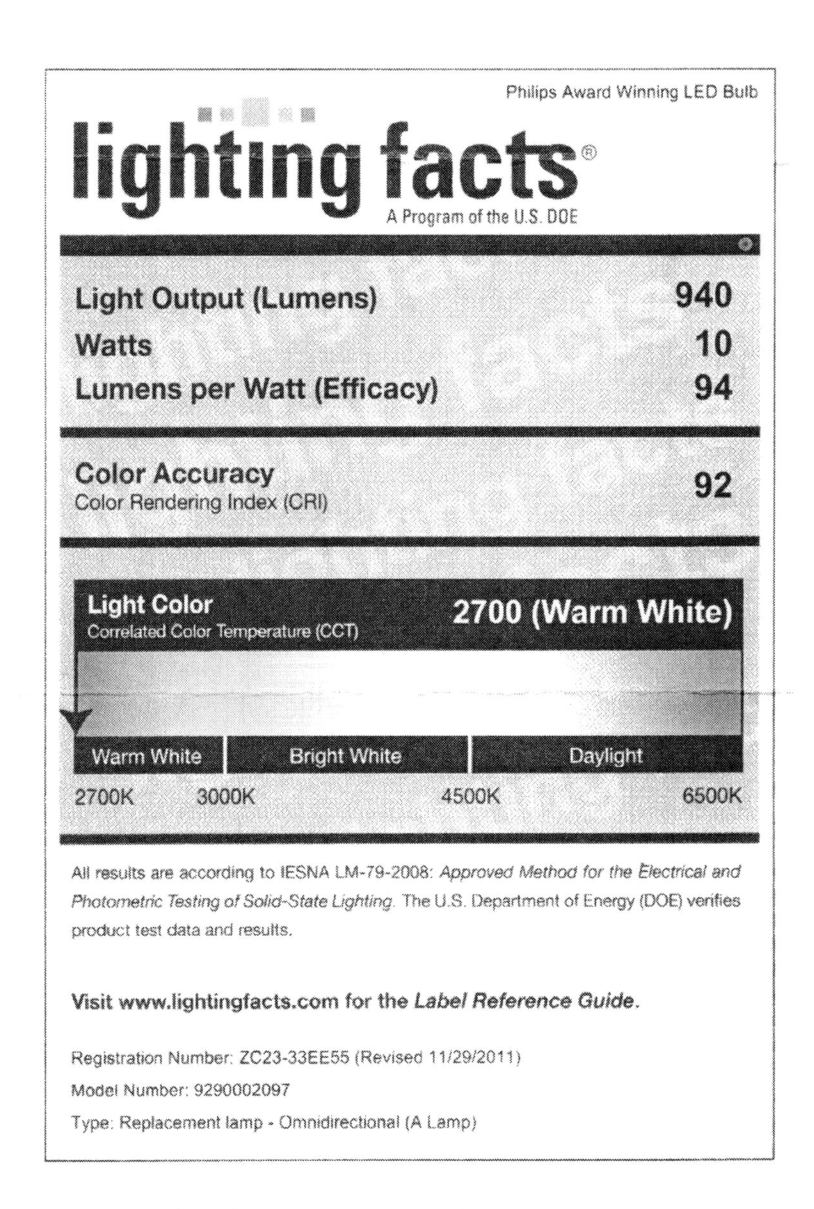

Figure 6-4. *Lighting Facts label—Philips L Prize winner bulb—new design*

7/Throw Out My CFL?

Perhaps at this point you are thinking that LED bulbs are the way to go. Maybe even that you should go out and buy some. Great...but if you're not buying new lighting fixtures at the same time, then you probably already have bulbs. What should you do with them?

The general rule of thumb when it comes to green living and prudent ownership is to finish using whatever you have before buying something new. For example, you might be excited to get rid of your old car and hop into a shiny new plug-in hybrid, but that car will have to be manufactured and shipped, and then something will have to happen with your old car. So while the new automobile will be more efficient than your previous one, the power and cost savings will still take a long time to accrue. Using this logic, there is no reason to rush the purchase unless you expect a drastic difference immediately (maybe you were driving around in an 18-wheeler?). This scenario also applies to lighting, as well as any other trade-off made for the sake of efficiency.

Theories aside, what if you just want to make your purchase now? If that's the case, you don't really need to think twice about your incandescents. They are quite cheap and they last for about 1,000 hours (about 11 months of life at 3 hours of usage a day), so if you can't remember installing them and they are in a room you use often, then they probably don't have much life left anyway.

CFLs are trickier because these bulbs aren't as cheap, they last much longer than incandescents, disposing of them properly is a pain, and they don't use much more power than a typical LED lamp. By way of example, I recently purchased a Philips AmbientLED bulb that I was going to put in my living room, in a fixture that I thought used an incandescent. I checked under the lampshade and saw that I was actually using a N:Vision CFL bulb. I tested the fixture at the socket, using a Kill A Watt meter, and found that it was running at 15W, a reasonable amount for a 60W equivalent bulb. I put in the Philips LED bulb and it consumed 12W of power. At a difference of 3W, I would have saved just $2.89 if I were to run the bulbs continuously for a year (assuming $0.11/kWh). At a more reasonable three hours a day, I would have saved myself about 36 cents over that same period.

Assuming a standard CFL costs about $2 today (they are generally sold in large packages, so they can be cheaper than this), it would take something like 5.5 years for me to get that $2 purchase back, assuming electricity prices don't increase. If I wanted to consider the added cost of the LED bulb (about

$26), then I'd be waiting a long time before the swap became cost effective. Even once I factor in the cost of the new CFL bulbs, I would need to buy to replace worn out ones during the life of the LED; case in point: it would take a number of years for the LED route to become cost effective.

Ultimately, it doesn't really make sense to throw out your CFL bulbs. If you are happy with the light they provide and you don't mind the startup delay (if they have one), then the main case for getting rid of them would be if they are in spots where you think they might break, like in a hard-to-reach ceiling fixture. Short of that they will do their job for years and barely cost you any more than an LED lamp.

8/Do LEDs Make Sense Today?

Whether or not LED bulbs make sense today is a judgment call that every person and business must make for themselves. A higher initial outlay has to be justified against future savings, both in terms of electricity saved and replacement costs. To some, it will seem obvious that LED bulbs are the way to go, while others will be happy to keep buying cheap incandescents as long as they are available. Stepping away from individual judgement calls, do LED bulbs make sense today?

There are three main factors to take into consideration before choosing to go the LED route: initial costs, power savings, and replacement costs. By tabulating these, every person, business, and organization should be able to figure out the the best option for them.

Initial cost is what holds most people and organizations back from buying a more expensive bulb. They see the incandescent bulb that has always worked for them, which is also the least expensive option, and the one that matches the rest of the bulbs they use, so it seems like the obvious solution. As CFL prices dropped, consumers were able to justify the extra expense thanks to their clear labeling about power savings and longer lives, but the mercury warnings have proven enough to turn many people off.

With today's LEDs, people see bulbs that range from $20–$40 and immediately write them off, knowing that the purchase of a single solid-state bulb would get them enough incandescents to illuminate their entire home for years. LED packaging clearly states the longevity and efficiency of the bulbs, but neither number is easy to understand. For instance, the 25,000-hour lifespan is not a term that is easy to take in and it's also one that can be hard to believe, especially by people who were burned by short-lived CFLs. Even if people comprehend the length of 25,000 hours (about 8.5 years of life if you use a bulb for 8 hours a day, 365 days a year), who is thinking that far into the future? The average person in the United States is said to move every 5 years...are they going to take their bulbs with them when they go? The prospect of long-term savings is easily defeated by short-term desire to skip a considerable expenditure.

First-generation LED bulbs might have done more harm than good for the industry, at least for the consumer segment of the market. These products, as previously mentioned, were expensive, were often not dimmable, and had blueish lighting that made people feel as if they were in a hospital or government building. These bulbs had the unfortunate effect of turning people off from a new category of product that just happened to have been released too early for mainstream buyers. Even the people that bought and liked them saw the rapid progress of LED bulbs and might wish they had waited another year or two before making the investment—especially considering that their bulbs had years of life left in them.

The next big factor in a purchase is power savings. It's one thing to read that LEDs use less energy than the competition, but the savings generated is much less easy to comprehend. How much will the total energy savings derived from efficient lighting actually mean when it's in a bill combined with your refrigerator, dryer, and air conditioner? Your bill could be a few cents less each month, maybe a few dollars if you have a larger home, but it's probably not going to be enough to make a considerable change. This is different for a business with dozens or hundreds of lights that run continuously, but even those savings will have to accrue over time before the bulbs buy themselves back.

The packaging and literature provided with LED bulbs often explains the savings over the life of a bulb given a set price of power (say, $0.11 per kilowatt hour) which can often be over $100, but that number is easy to dismiss. A yearly savings amount or buy back timeframe (the point at which an LED bulb will become the cheaper option than an incandescent would be) is helpful, but it's difficult to convince people about savings when the initial price is higher. After all, casual consumers just want to decrease their initial bill and savvy ones fully understand the opportunity cost of investing that money in expensive lighting instead of something else. Regardless, the purchase does make sense from a power savings standpoint, it's just a matter of whether you will keep the bulbs long enough that they will be able to give you a worthwhile return on investment.

Lastly, there is the often forgotten matter of replacement costs. This factors in not only the life of the bulbs (considering that you might need ten incandescents or three CFLs to match the life of an LED bulb), but also that bulbs are often in inconvenient places. If you need to rent a ladder or pay a person to switch a bulb, that cost should be taken into consideration. This is often more of a factor for businesses, but people with high ceilings would do well to take it into account.

Using these factors, it should be relatively straightforward to determine whether LED bulbs make sense for you or your business. Many people don't like to mix bulbs, but upgrading from room to room might make sense. Selective upgrading might be worthwhile as well—just changing the lights that you use the most or that you leave on the longest, so that you can save money without spending hundreds on bulbs. Regardless of your methodology, it's worth understanding that today's LED bulbs save money over the long run and they are a big step up from what was offered just a few years ago. That doesn't mean they are the correct choice for everyone or under all circumstances, but they are getting closer and closer to that point.

9/LED Lighting Today and Tomorrow

The future of the LED is bulb is, to some extent, already determined. If Haitz's Law holds true, the technology will become increasingly affordable and, following that, the lamps will become ever more popular. With clear advantages over both incandescent and CFL lighting, LEDs will likely be extremely widespread in four to five years and the other two will continue to wane. The competitors won't ever go away—for example, Edison and Marconi-style bulbs are available in delightfully old-timey versions—but they will be increasingly relegated to specialty uses, in favor of more efficient technology.

A commonly held belief among people in the industry is that the explosion in growth will come when LED bulbs reach a certain price point. Some insiders put this at about $14, while other accounts (*http://nyti.ms/ql2jTp*) predict $10, or even $5. Still, others regard this watershed level as a closely guarded secret and refuse to talk about it. Regardless of the amount, this will mark an inflection point when consumers will begin to clear the mental hurdle of solid state lighting's high initial price point and understand that it is a more affordable option. It's going to be a revolution in the consumer lighting industry, even if efficiency of scale prevents this level of pricing to arrive at all manufacturers at the same time.

When the price of CFLs hit about $14, their popularity exploded and mass consumer adoption started. While there is no guarantee that this will happen with LED lighting, growth will start as prices drop, and we can reasonably expect that a sharp increase will occur as bulbs drop below the point that the mainstream consumer finds to be unpalatable.

CFLs experienced a sharp increase in sales in 2007 and reached their peak sales figure that year (*http://bit.ly/NeRy7Z*). Their sales have been steadily declining since the early part of that year and show no signs of improving. Incandescents experienced a small upswing in sales in mid-2010, according to the National Electrical Manufacturers Association's (NEMA) July 2011 numbers. This 1% rise gave incandescents a dominant 79% market share—though the report does not seem to take LEDs into consideration at all. Of course, part of the drop in CFLs was due to their price during hard economic conditions after 2007 (*http://nyti.ms/oBhPMR*).

Despite LED sales not yet exploding, hopes are high for the future. In other words, LEDs won't be left out of reports for much longer. According to recent DisplaySearch numbers (*http://bit.ly/qSWeGn*), LED lighting usage increased 20% in 2010 and is on the rise. Their figures put LED lighting usage at 1.4% in 2010, and the firm expects it to rise to 9.3% in 2014. Government policies, such as the use of LEDs in street lights, could accelerate this trend. Japan is the largest user of LED bulbs, having an incredible 63% of the market. Additionally, Pike Research estimates that by 2021 LED lamps will have over 50% share of the commercial market (*http://bit.ly/t5uLaR*). Over the next 10 years, the share of incandescents and fluorescents will drop thanks to SSL, but the report also notes an interesting side effect: the long lives of these lamps will cause a "steady decline in the commercial lamp market."

One step forward in the movement towards mass LED sales was Home Depot's release of a sub-$20 LED bulb in August 2010 (*http://nyti.ms/bkZLi1*). The Lighting Science Group's EcoSmart LED A19 bulb was a 40W equivalent (452 lumens). It sold for $19.97, and Home Depot reportedly had trouble keeping them in stock when they were first released because of high demand. Now, less than two years later, that same bulb is still available from Home Depot, but it sells for just $9.97. It's a clear display of what the technology and scale can do to a growing market, even if it is just one bulb. (That said, it is an in-store exclusive, so we have no way of knowing if it is a loss leader for the company.)

Over time market forces will almost inevitably make LED bulbs the most popular choice, but before then, other factors are putting their thumbs on the scale. The Bright Tomorrow Lighting Prize (better known as the L Prize), held by the US Department of Energy, was designed to promote the usage of solid-state lighting as an alternative to incandescents. The contest consisted of two awards in order to facilitate the replacement the standard 60W incandescent and the PAR 38 halogen incandescent bulb (i.e., indoor flood lights). There is a third prize, but it has yet to be fully defined.

The first prize, for the 60W bulb replacement, was won by Philips Lighting North America in August 2011 with the 10A19/LPRIZE-PRO/2700—essentially, a high-performance version of the AmbientLED. This outcome (*http://bit.ly/N4mHcp*) was something exciting for LED insiders. To win, the bulb had to beat five separate components of the L Prize requirements, including lumen level (over 900 lm), wattage (under 10W), efficacy (over 90 lm/W), CCT (within the range of 2700K–3000K), and finally, an adequate CRI (over 90). The Philips bulb passed with 910 lm, 9.7W, 93.4 lm/W, 2727K, and a CRI rating of 92. Over 200 units of that bulb passed the LM-79 test procedure before the award was granted.

While the prize has been awarded already, the longevity testing is still underway. The 200 bulbs are being run continuously at 45° Celsius, and the data for the first 7,000 hours was used to extrapolate (at 95% confidence) that the bulbs would last the expected 25,000 hours at adequate brightness levels. The testing put the model's survival rating at 99.3%, while the prize's requirement called for a mere 70%.

Philips pulled in $10 million for the win, but the the real prize was the improvements that were made. For example, at the time that Philips took home the prize, the company already had a 60W equivalent bulb for sale, the EnduraLED, that ran at 12.3W. Meanwhile, the L Prize-winning design runs at 9.7W, making it over 25% more efficient. The L Prize winner is also brighter, running at over 900 lm while the EnduraLED ran at 800 lumens.

The PAR 38 part of the contest was closed in January 2011 and was reopened (*http://bit.ly/MpimWJ*) in March 2012. To win the $5 million prize, a company must produce 32 samples of a PAR 38 halogen incandescent replacement lamp that emits over 1350 lm, consumes 11W or less, has a luminous efficacy of 123 lm/W, has a CCT of 2750K or 3000K, fit the industry standard size/shape, and has a CRI of 90 or higher (the R9 value must be 50 or higher). Finally, the lamp must maintain 70% its initial brightness (L70) for 25,000 hours.

While the L Prize is notable, the cash prize and bragging rights are small motivators compared to what will come as prices drop and millions of lamps are sold. This is what will truly propel companies and enable them to take market share away from incandescents and CFLs, but competition can still slow down the LED's path to dominance. A US Congress bill mandated that certain incandescents would be effectively outlawed in 2012, thanks to restrictions on power usage (more on this later), but since the bill was passed there have been a number of advances in incandescent technology. Now bulbs like the Philips Halogena Energy Saver are able to decrease power usage by up to 30% compared to older designs. These bulbs are more expensive than traditional incandescents, and they are less efficient than CFLs, but they are very affordable. Unsurprisingly, they are an attractive alternative for consumers.

An upcoming competitor in the updated incandescent space is a company called Vu1, which uses a technology known as electron stimulated luminescence (ESL) in their lamps. This allows for instant-on, mercury-free lighting, but isn't quite as energy efficient as CFLs. The technology ESL uses is not unlike that of the cathode ray tube (CRT)—the large, boxy televisions and computer displays of the past. Until LED bulb prices drop further, ESL could be a competitive solution, provided that products are readily available and consumers continue to shy away from CFLs. ESL's strengths include that the shape of the bulb is quite malleable (unlike fluorescents, which require a tube, and thusly a helical shape to save space), the color of the light is close to that of incandescents, and less heat production than incandescents. Best of all, they should eventually be as cheap as CFLs to produce.

Vu1's 60W equivalent, the R30, operates at 2800K (warm white), has a CRI of 85, puts out 600 lumens, and sells for just $14.95. The bulb has a life of

11,000 hours and operates at 19.5W. It works in recessed spaces, making it a reasonable replacement bulb for standard incandescents. The bulb has a longer life than a typical CFL, but at this point, consumes more electricity and costs more.

As for incandescents, researchers say that more efficiency gains are on the way, so they won't fade out as quickly as we might expect (*http://nyti.ms/9mV9n*). Low prices and cost-sensitive consumers will ensure that incandescents stick around for the time being, but there will be limitations on their future. Barring further advances they are still less efficient (in the sense of lumens-per-watt) than LED bulbs, and while they are more affordable now, the price of LEDs should continue to drop and present a cost-effective alternative.

Technology isn't the only force pushing away incandescent bulbs. In a number of countries, including the United States, Australia, the United Kingdom, and China, governments are intervening and getting ahead of the market, putting legislation in place that phases out, and ultimately bans, high-power consumption lighting. Of course, a ban is never the goal—it's just the enforcement measure needed to improve lighting performance.

In the United States, the 2007 Energy Independence and Security Act (*http://1.usa.gov/tB3WGk*) (EISA, also known as the Clean Energy Act) put new efficiency standards in place for both appliances and lighting. The bill didn't ban incandescents outright; rather, it required greater efficiency from them, making it so that the bulbs that were on the market in 2007 would not be able to be sold in 2012. The mandate for about 25% greater efficiency meant new incandescents would have to be developed, or that buyers would have to seek out alternatives.

Specialty bulbs would be exempt, but a main target was the 100W incandescent. The bill said that bulbs with a lumen range of 1490-2600 must have a minimum life of 1,000 hours and use no more than 72 watts by January 1, 2012, which effectively outlawed the present-day 100W bulb. Next would come a restriction on bulbs at 1050-1489 lumens, which would not be able to exceed 53W after January 1, 2013. 40W and 60W bulbs would not be affected by the legislation until 2014, as per the set levels (750-1049 lumens and a maximum of 49W).

Unsurprisingly, today you can purchase a Philips EcoVantage 72 watt bulb (it's a halogen lamp, a type of incandescent) for under $7 for a two-pack. The packaging clearly states "72W = 100W" which explains how it consumes less power but is just as bright. The clear (non-frosted) version consumes 72W, produces 1,490 lumens, and has an averaged expected life of a the government-minimum 1,000 hours. The company's Ecovantage explanation page clearly states that the bulb is EISA 2007-compliant and that "compared to a 100W Natural Light A19 incandescent bulb with 1,350 lumens, the 72W EcoVantage bulb with 1,200 lumens provides similar light and at least 28% energy savings."

Philips also has a 29W EcoVantage model that serves as their 40W equivalent and a 43W bulb that is their 60W equivalent, rounding out their offerings.

Ecovantage or not, LEDs will win in the end. While legislation will help, technology will prove the deciding factor. LEDs will be so formidable that the lower price of incandescents will be a non-factor. This will take time, but as LED technology advances in the labs, it will trickle down and affordable bulbs will be better than ever. A recent prototype from LED manufacturer Cree, known as the 21st Century Lamp, was able to produce 1300 lumens at an incredible 152 lm/W, a number far beyond Philips' L Prize winner. Also from Cree is the newly announced XB-D LED (*http://bit.ly/xqdwpH*). It might not sound like anything special, and ultimately it might just turn out to be another light-emitting diode, but this tiny unit is able to provide over 135 lm/W in cool white (6000K) and over 105 lumens per watt in a more indoor-friendly shade of white (3000K). In their labs Cree has an LED capable of 254 lm/W (*http://bit.ly/HESkLK*), so we know that incredible things are in the pipeline

Of course Cree's 21s Century Lamp and the projects hidden away in their labs are prototypes that will probably never be available for sale. It's little surprise that a new year brings new LEDs, but the important thing is that the technology is out there. As prices drop, we'll start to see the tech and high-end parts trickle down, and soon enough LED bulbs will replace 75W and then 100W incandescents. By that time, 40W and 60W replacement bulbs will be more affordable than ever and the LED will be well on its way to being the world's dominant lighting technology.

One curious product that we probably won't be seeing much of in the future is the Panasonic LDAHV4L27CGLED Nostalgic Clear Type (*http://bit.ly/HESkLK*). This is an LED bulb designed to look like an incandescent. It has all the features and perks of an LED bulb, but when it's off it appears to have a filament inside. It lacks a heat sink, so it's only good for a paltry 210 lumens, or 20W equivalent. It's a strange and silly product that assumes people won't want one of those new-fangled solid-state lights or at least that they are OK with a dark house and fooling themselves into thinking they are using an old-fashioned bulb. Ultimately, it's a handsome bulb...but what's the point?

10/The Future

It's nice to think that the best technology always triumphs, but we know that's not the case. The QWERTY keyboard layout has been maligned for years, yet it's so ingrained into our minds (and laptops) that we can be relatively sure it's not going anywhere. And, of course, the most often cited example of a better technology being left behind is Betamax's loss to VHS during the so-called Video Format War (*http://bit.ly/bF9o4V*) of the mid-1970s. Ultimately, it's unlikely the LEDs will lose out to another technology, especially incandescents, but it will be a drawn out fight and hard-won victory.

As prices drop and consumer awareness grows, LEDs will flourish. Boosters like the L Prize, government regulation, power company subsidies, and (perhaps) rising energy prices will all assist in LED adoption. The case for increasingly efficient lighting is inarguable, and LEDs have the best chance right now, especially considering the trajectory of the industry and the downward trend of the competition. As power becomes an issue for everyone, awareness will grow and the huge gains in efficiency found in solid-state lighting will be welcomed as the lowest-hanging fruit possible, saving individuals, countries, and the world billions of dollars, not to mention tons of carbon dioxide, a year.

It's fun to think in terms of winning and losing, but things are rarely that simple. The incandescent will probably always be available in some form, and while CFLs are largely recognized as a transitional technology, they'll be around for some time before they disappear. The better way to think of the situation is as progress: Solid-state lights are tools for a digital age, just as incandescents were the tools for an analog one.

While that's a convenient observation more than it's a reason for success, it will be a strong motivator for buyers. Everything in our lives has gotten smarter—from our phones to our televisions to our refrigerators—and now the hold-outs, like our lamps and door knobs, are finally catching up. It's not solely the gadget-ization of the world around us, it's also the optimization of costly, inefficient technology that needs to change with the times.

One of the conclusions that this leads us to is just how different solid-state lighting is from what most of us grew up using. LED lamps are solid-state. They have integrated circuits. They last for decades. They can survive a bump, even when hot.

LED replacement bulbs are something new, but if you were to look at their marketing materials and associated packaging, you'd see that the industry wants little more than to position them as a replacement for something old.

In fact, to the casual observer, they just seem like a longer-lived, more expensive, futuristic-looking version of the incandescent bulbs on the same aisle. And while the end result of the two products is the same, that thinking couldn't be further from the truth.

The industry has largely taken it for granted that consumers fear change and don't want something new, when that's exactly what LED manufacturers could offer them. This replacement of a 100+ year old technology is offering something completely new, while manufacturers are trying to bill it as a better version of the same old thing.

In the same way that the mobile phone can't compete with a landline (it needs to be recharged and it drops calls?), and a tablet can't compete with a notepad (there's no pen and it doesn't work in the sun?), the LED bulb should not attempt to compete with the incandescent...except in the most conventional of terms. Yes, they both solve the same problem, but this is a genuine shift in the method of reaching that end point—the production of light—that should be understood and embraced.

When real, transformative technology comes along, it may not be immediately acceptable. Potential buyers may balk at the prospect of doing things differently, but that hesitancy is not overcome by competing head-on with older technology. The newcomer can't compete against something ingrained; rather, it must change the conversation. Significant technology changes—like the wireless telegraph, the electric car, and the LED bulb—are replacements, not upgrades.

Telling consumers how much they'll save over the next 22.5 years isn't ever going to be a persuasive case. Twenty dollars today is a lot more useful than a few pennies saved in a power bill each month for the next few years, so a new argument needs to be made. The argument for "better" light is out the door—nothing will ever match the quality of sunlight or the glow of an incandescent that we've long understood to be real indoor light—so where does that leave the LED?

It's time to embrace technology and see it as a replacement, not a power-saving alternative to something we'd rather have—like some sad, low-fat variety that we know is better for us, but that we don't really want. Building on the LED's technology is surely the most effective way to differentiate it and go someplace where incandescents, however efficient, cannot follow.

And what does building on technology mean? It's not just about improving efficiency or lowering prices—those are just improvements on the same light bulb that's been in use since the late 1800s. Real technological progress is networked controls, power over ethernet (PoE), maybe even wireless power. Any sufficiently advanced technology, to borrow a phrase from Arthur C.

Clarke, could be the sort of breakthrough that puts modern lighting on people's radar. It might not directly lead to the sale of millions of units, but it would certainly make headlines and draw attention to the improvements the solid-state lighting makes possible.

Why should technology even go into a light? This is pretty simple: lights are properly placed for a number of duties and they already have power running to them. This means that the cost of integrating technology into a light is cheaper than running power lines and installing an entirely new device. Furthermore, although we've been focused mostly on replacement bulbs that use LEDs, the future will not be limited to these. LED fixtures, sometimes known as luminaires, will be an important option and these larger fixtures won't be limited to use in a socket. More importantly, fixtures are big, often expensive things, so there is no pressure to get them to the sub-$5 point. This means the integration of technology can make perfect sense, though it might not ever be in a lamp designed to light a living room. LSG's Glimpse (*http://bit.ly/KWtyZX*) is a prime example of a fixture that can work in a retrofit situation. It's Energy Star-compliant, dimmer-compatible, offers up to 750 lumens, and is designed to last 50,000 hours. The Glimpse retrofit downlight (6-inch) is available today at Home Depot and sells for $37.

While wirelessly powered lamps would be a major step forward, it'll be more feasible to have bulbs that are wirelessly controlled. At the 2012 Consumer Electrics Show a company called Fujikom had a Z-Wave wireless light bulb on display known as the LeDenQ (*http://bit.ly/y5eH8W*). It is an LED lamp that produces 880 lumens and is rated at 40,000 hours, but that's just the start. It can produce around 20,000 different colors and its wireless controller will allow owners to change the bulb's settings. This will be possible using an app or the included remote control. It might sound like a parlor trick—and the bulb isn't truly wireless, it still needs power from the socket—and Z-wave isn't anything new (it's just another wireless home automation solution), but it's a step in the right direction.

IP-addressable lights are another logical step forward. They connect through a wireless or powerline network and can then be controlled and monitored remotely. These might not replace the humble light switch anytime soon, but they certainly could. Before then, it would make controlling arrays of lights easier than ever and lights would be trackable, so customers could monitor usage or make sure they are getting their promised 25,000 hours. While these might sound like nice toys to have at home, just imagine if you were in charge of the lights at a factory or warehouse. Knowing which lights were not operational would be a huge task, but intelligent fixtures would make it simple. Networked lighting systems already exist, as do LED lighting systems that are powered by PoE and that have motion sensors so they only light up when they detect motion, but these are on the cutting edge today. In the future, these systems will be more widespread and bring convenience and power savings wherever they are installed.

The entire concept of intelligent lighting could be brought to market much easier by tying into a platform like Google's *Android@Home*. This would allow Android-powered devices to interact with a home's functions, including lighting, HVAC, and audio. Some companies are already trying out such forward-looking technologies, including the Lighting Science Group (*http://bit.ly/NeSaug*) (LSG). At Google I/O 2011, the search company's annual developer conference, LSG said they were working on lighting that could be controlled through a smartphone app. Flipping a lamp on and off is just the start: you could program your phone to tell your lights to dim when you are watching TV (using an NFC chip on your side table) or to turn on when you get home, using the phone's GPS for geo-awareness. This solution would use 6LoWPAN (IPv6 over low-power WPAN) to make your bulbs part of the so-called "Internet of Things."

Another option is the NetLED (*http://bit.ly/MOvGCO*), a fixture designed to replace fluorescent tubes, that interacts with an app and remote servers for control. At over 60,000 yen (about $740) the kit isn't cheap, but the lights can be controlled by your smartphone or tablet via their dedicated wireless router. The app offers on/off controls as well as dimming, preset lighting patterns, and charting of usage data.

Audio company Klipsch combined speakers with LED lights in their $600 LightSpeaker In-Ceiling Lighting and Audio System (*http://bit.ly/m7JfE4*). Perhaps it wasn't the company's most successful product, but it combined lights with wireless speakers, making for a convenient system that worked with up to eight speakers and pumped out 20W of audio power per speaker... without running a single wire. The speaker-light combo fit in a standard recessed "can" fixture, which wouldn't have been possible without the size, efficiency, and flexibility of the LED.

Of course, if our look into LED lighting has taught us anything, it's that consumers are extremely sensitive to pricing. Integrating IEEE 802.15.4 wireless into a lamp, as LSG's experiment calls for, apparently wouldn't make for a larger bulb, but it would make for a more expensive one. So far, manufacturers have just about been able to convince businesses and organizational buyers that new lamps are worth the investment, but consumers, with lower power bills and increased price sensitivity, are harder to win over.

Perhaps if their wallets can't be appealed to, their gadget lust can be? It is likely that the lighting market will diverge in the future, with part of it continuing to cut costs as quickly as possible to get back to that sub-$5 bulb. The other part of the market will take full advantage of solid-state lighting's qualities and go the intelligent route, using intelligent products like those offered by Redwood Systems. After all, a light in a parking garage that can tell the operator what capacity the garage is at, turn on when it detects motion, or have a built in security camera is extremely useful. And the best part is that such a setup will also save money in the long run.

Advances, gimmicks, and gadget-lust aside, the LED has an exciting future ahead. The lighting world is ready for a revolution, and it's probably not going to be incandescents or CFLs that make it happen. Over the next few years (it's going to take some time for all of today's CFLs to burn out), there will be rapid changes in lighting. This might seem like an optimistic statement, but we are reaching a watershed moment, not unlike that of the advent of the mobile phone or broadband Internet. Something useful, affordable, and clearly better is about to be widely available and very much on the radar of businesses and consumers. It's just a matter of time before the LED is how our world is lit.

When this will happen is another matter. Most people will avoid the question completely, while others will tell you "soon" or "over the next few years." The likelihood is that there will be a rapid uptake in LED lighting when certain conditions are met. If there were to be a major rise in the price of power then adoption would almost certainly increase, though that's not something we tend to see fluctuate enough that it's on the average person's radar. Another condition could be that power companies simply start to give out LED bulbs, as they did with CFLs. They will want consumers to save energy and free, high-efficiency bulbs are an ideal route.

Of course, the final and most definitive path to LED bulb adoption will be their inevitable drop in price. As noted in the last chapter, numbers vary on this, but somewhere between $10 and $15 per bulb is where many people forecast that LED marketshare will grow dramatically. This might vary based on economic conditions and how many high-wattage incandescents people have squirreled away in their garages, but it's a reasonable estimate for the point where people will no longer be able to ignore the advantages of solid-state lighting. We're already starting to see LED adoption in street lamps, hotels, and other near-continuous-use situations; it's the consumer sector that's lagging. These are the buyers that are the most resistant to the initial investment. As prices drop, it's inevitable that adoption will increase...especially once all those incandescents people have hoarded start to burn out!

And while the focus is on the future, it's worth remembering that the downward slope of pricing is in full effect today as well. Excellent LED lamps are available for under $30, and fully adequate ones can be purchased for under $20. In February 2012, a company called Lemnis announced their Pharox Blu series of affordable LED bulbs. The Pharox Blu 200 retails for just $4.95 while the Blu 300 is $6.95! Sure, they produce just 240 and 360 lumens respectively, they are rated at 15,000 hours, they aren't dimmable, and have sub-90 CRI ratings, but it's hard to argue with those prices. In a press release, the company noted that it saw a 500% growth in sales directly after lowering their pricing to these levels (*http://prn.to/xkDxv5*). These might not be the best bulbs on the market, but it's a sign of things to come.

To put some numbers behind this growth, a recent estimate by a market research firm estimates that the LED replacement lamp market will grow from a value of $2.2 billion in 2011 to $3.7 billion in 2016, with the growth being led by China. Furthermore, they expect average prices to fall by 14% a year during the same period (*http://bit.ly/MxWtnJ*). While that's just one set of numbers from one firm (albeit one that's been studying the photonics market since 1979), it's indicative of the incredible growth expected in the industry. The push for energy efficiency, the desire for green solutions, the appreciation of quality light, and a gentle nudge from government agencies will all combine to push the LED lamp from a new technology to the standard.

The Department of Energy regularly publishes technical roadmaps (*http://bit.ly/NgowTj*) that explain their future outlook. These are technical documents, but they are endlessly interesting if you are curious about where experts believe solid-state lighting will be in the next few years. The latest roadmap document, from January 2012, had set targets for up to 2020 with goals that are "challenging but achievable." These focus on the technical gains that should be achieved in the near future, while the organization's Manufacturing Roadmaps are concerned mainly with pricing. From 2010 to 2020, the latest roadmap (July 2011 (*http://bit.ly/neonF9*)) expects the cost of the LEDs in a lamp to fall from 50% of the total cost to less than 25%, and the relative manufacturing cost of a lamp to be approximately 10% of what is was in 2010. Specific numbers aside, the roadmaps project tremendous gains that will greatly reduce the cost of solid-state lighting and increase its accessibility for all buyers.

The growth of the LED industry and of LED lighting is, in many ways, a logical extension of the growth of the semiconductor industry. Over the years we've seen a clear trend: where semiconductors go, they take over. Whether it's for communications, data storage, digital imaging, optoelectronics, or any other number of other aspects of our lives, semiconductors are just too efficient, and improve at too rapid a pace, to be ignored. Incandescent lighting had a long and prosperous run, and it will probably never go away entirely, but we have found something better; it's just a matter of time until the light-emitting diode is the tool by which we illuminate the world around us.

A/DIY LEDs

One of the best ways to learn about LEDs is by tinkering with them. While encased in lamps, LEDs might be tiny, sensitive, and hard to understand, but when out in the wild they are affordable, easy to use, and rather simple devices. This makes them ideal for experimentation and a great way to learn the basics of electronics.

The previous mention of do-it-yourself LEDs covered the Throwie, the simplest of the LED circuits. It combines a battery and an LED into a easy to understand device that anyone can make in minutes. The next step up from that two-part project is adding a resistor to the mix. This will allow you to adjust your circuit so that the proper amount of power is going to the LEDs which will, in turn, ensure they get the maximum lifetime possible as well as the optimum brightness. Without a resistor, it's easy to send too much power to the LEDs and shorten their lives significantly, or simply break them.

To figure out the resistor you need in your LED circuit, some basic math will be required. The formula is based on Ohm's law, where V = voltage across the resistor, I = the current through the resistor, and R is the resistor value. Here are three ways of stating Ohm's law:

```
I = V/R
V = IR
R = V/I
```

It's this last one (R = V/I) we want to use:

```
Resistor rating = (battery voltage - LED forward voltage) / LED forward
current
```

This means you need to know the voltage of your power supply, the forward voltage of your LED, and the forward current in amperes (amps) that the LED draws. Both the forward voltage and forward current will be provided by the LED manufacturer (if you bought the LED from a retailer, they may include the specs along with the packaging or have a link to the manufacturer's data sheet on their website).

The forward current for LEDs are specified in milliamperes (mA), so you'll have to divide your mA number by 1000 in order to get the amount of amps.

A common red LED has a forward voltage of around 2.0 volts (V), and a forward current of 20 mA. So the math for a single red LED powered by a 9 volt battery would be calculated like so:

```
(9.0V - 2.0V) / (20 mA/1000) = 350 ohms
```

Because people tend to opt for the nearest higher rated resistor, you could opt for a 390 Ohm one (the color code would be Orange White Brown), but 220 ohm resistors are more common, so you could put two of them in series for 440 ohm. Then, try the same circuit with a single 1 kilohm resistor. You should see the LED get less bright than with the two 220 ohm resistors. Figure A-1 shows a superbright green LED with a 1K resistor.

This LED circuit is shown on a *solderless breadboard*, which allows you to quickly wire up components. The rows and columns are connected to each other in such a way that the resistor and left leg of the LED are tied together, and the yellow wire and the right leg of the LED are tied together. Similarly, the resistor is connected to the red wire, and the yellow wire is connected to the black wire. If you trace the circuit starting from the red wire (the battery's positive lead), the circuit travels in this path: from the positive lead, through the resistor, then through the LED, and back into the negative (black) lead by way of the yellow wire.

Because the superbright has higher forward voltage (around 3V) and higher forward current (around 80 mA) than a red, it would only need a 100 ohm resistor. However, if I did that, it would be too bright to take a decent picture, hence the 1K resistor.

But what if you wanted two LEDs in a series? That's entirely possible, you'd just need to adjust the formula. In this case, the formula will change to:

```
(9.0V - 2.0V - 2.0V) / (20/1000) = 250 ohms
```

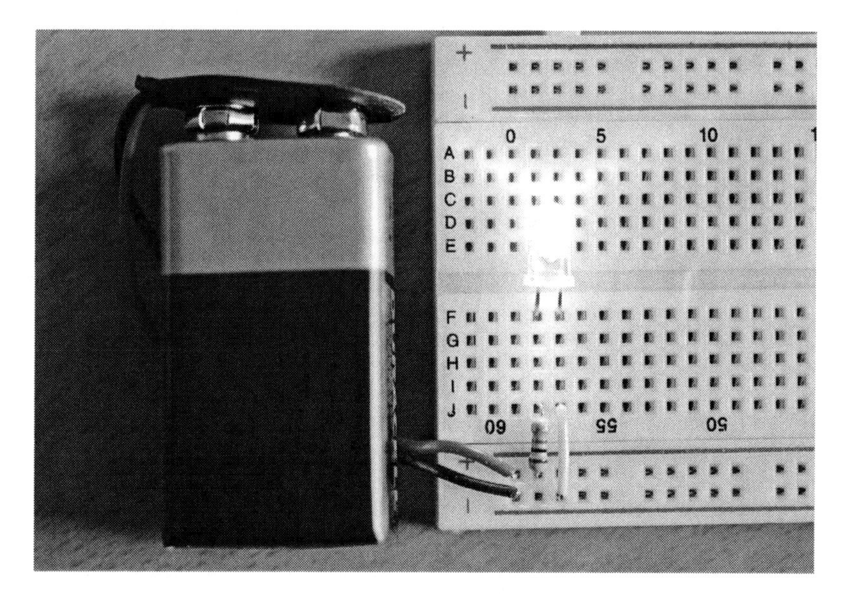

Figure A-1. *LED circuit*

Twice the LEDs means twice the voltage is required, so you'll have to change your resistor in order to compensate for the difference. Pretty easy, right? Figure A-2 shows the circuit wired up.

But what if you want to test out the deleterious effects of overvolting an LED? The math makes it clear that this is pretty easy to do, you just need to offer too much power to the LEDs or not enough resistance. Figure A-3 shows a fried LED.

If you decide to try frying an LED, don't hold it in your hand, and do so in a well-ventilated area. One way to avoid holding it in your hand is to use a breadboard to hold the LED and connect the battery further down the breadboard. You can safely throw the LED away after it's cooled down.

By playing with the resistance variable you'll be able to see the LED in its full spectrum—from dim, to bright, to peak brightness, to burnt out. The LED's lifetime will travel along this same asymptotic curve if you were to plot it out, with there being an inverse relationship between brightness and lifetime.

If you're inclined the explore more hands-on experiments with LEDs, check out *MAKE: Electronics* (*http://oreil.ly/GGh3xB*) by Charles Platt (O'Reilly/MAKE).

Figure A-2. *Two LEDs wired on a breadboard*

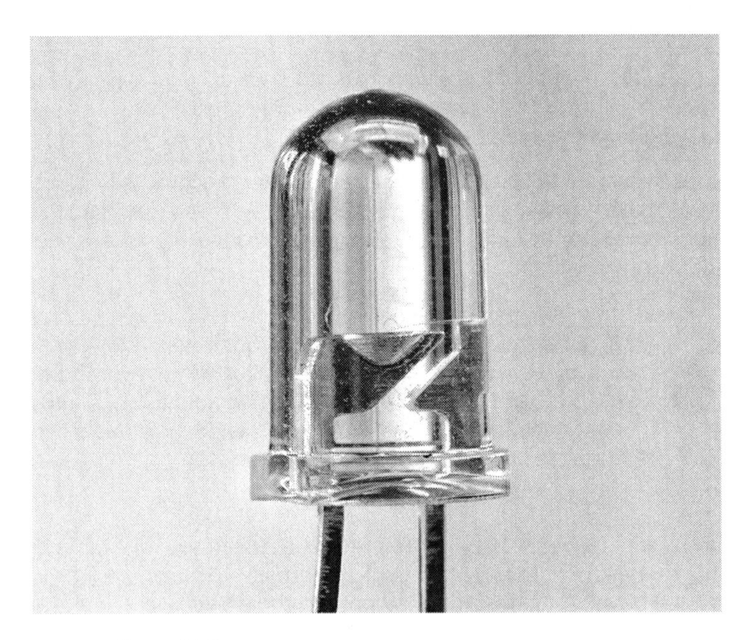

Figure A-3. *5mm LED*

About the Author

Sal Cangeloso is a lifelong technology enthusiast who founded computer hardware site XYZcomputing.com in 2003. In 2007, he started working for Geek.com. He now runs Geek.com and ExtremeTech.com. Sal is an inquisitive guy who loves computers, cameras, and pretty much anything electronic.

Built with Atlas. O'Reilly Media, Inc., 2012.

Have it your way.

O'Reilly eBooks

- Lifetime access to the book when you buy through oreilly.com
- Provided in up to four DRM-free file formats, for use on the devices of your choice: PDF, .epub, Kindle-compatible .mobi, and Android .apk
- Fully searchable, with copy-and-paste and print functionality
- Alerts when files are updated with corrections and additions

oreilly.com/ebooks/

Safari Books Online

- Access the contents and quickly search over 7000 books on technology, business, and certification guides
- Learn from expert video tutorials, and explore thousands of hours of video on technology and design topics
- Download whole books or chapters in PDF format, at no extra cost, to print or read on the go
- Get early access to books as they're being written
- Interact directly with authors of upcoming books
- Save up to 35% on O'Reilly print books

See the complete Safari Library at safari.oreilly.com

Spreading the knowledge of innovators. oreilly.com

CPSIA information can be obtained at www.ICGtesting.com
Printed in the USA
BVOW081756190712

295711BV00001B/2/P